THE ORIGIN OF LIFE

A SERIES OF BOOKS IN BIOLOGY

Editor: Cedric I. Davern

THE
ORIGIN
OF
LIFE

A Warm Little Pond

CLAIR EDWIN FOLSOME

UNIVERSITY OF HAWAII AT MANOA

W. H. FREEMAN AND COMPANY

San Francisco

Cover photograph on softcover edition: The Crab nebula. This nebula is the gaseous and material debris of a supernova that is now 6 light years in diameter and expanding at the rate of 5 million km/hr. The core of the one-time star is now compressed into a neutron star (a pulsar), which cannot be seen in this photograph. In AD 1054 Chinese astronomers recorded a new star so bright it could be seen during the day; after about a year it faded from sight. (Copyright by the California Institute of Technology and the Carnegie Institution of Washington. Reproduced by permission from the Hale Observatories.)

Sponsoring Editor: Arthur C. Bartlett. Project Editor: Patricia Brewer.
Copyeditor: Patrick J. Foley. Designer: Marie Carluccio.
Illustration Coordinator: Cheryl Nufer. Production Coordinator: Linda Jupiter.
Artist: Georg Klatt. Compositor: Graphic Typesetting Service.
Printer and Binder: The Maple-Vail Book Manufacturing Group.

Library of Congress Cataloging in Publication Data

Folsome, Clair Edwin, 1935–
    The origin of life.

    (A Series of books in biology)
    Bibliography: p.
    Includes index.
    1. Life—Origin. I. Title.
QH325.F57   577   78-10809
ISBN 0-7167-0294-0
ISBN 0-7167-0293-2 pbk.

Printed in the United States of America

9 8 7 6 5 4 3 2 1

To Jo, and all the other giants

*I*t is often said that all the conditions for the first production of a living organism are now present, which could ever have been present. But if (and oh what a big if) we could conceive in some warm little pond, with all sorts of ammonia and phosphoric salts, light, heat, electricity, etc., present, that a protein compound was chemically formed ready to undergo still more complex changes, at the present day such matter would be instantly devoured, or absorbed, which would not have been the case before living creatures were formed.

CHARLES DARWIN

CONTENTS

PREFACE

This is a short record of a long adventure, which began at the dawn of awareness and which will see no end. Its object is to be able to explain—perhaps to re-create—the origin of life on this planet. A vital component of this goal is also to know enough about the phenomenon of life to become able to predict whether life is rare or unique or whether it is a common and necessary feature of the Universe.

Our history, literature, and folklore are rich with the wonder posed by these ideas. Only recently—over the past century or less—have our science and technology yielded enough raw data for us to develop our knowledge and sense of wonder to form useful theories about the nature of the Universe and our place within it. Astronomy, physics, and chemistry have pooled spectroscopic techniques, theories, and data showing that the chemical composition of the solar system is in fact universal. Nuclear physics and astronomy permit us to set a time scale for the evolution of stars and the ages of rocks and meteorites.

Astronomy, physical chemistry, and geology combine to offer a realistic view of the way planets form, evolve, and develop atmospheres. Chemistry and biology set the material stage for the origin of life. Philosophy, too, is a vital component. An objective frame of reference, based on observation and experiment, permits us to begin to answer origin questions. This can be phrased as: "a reasonable explanation exists for every event." Some explanations are now more reasonable than others, and some events are now inexplicable observations. This is the fascination of science.

This book is for the curious person who has an interest in science and in our local cosmology. It seeks to explain my view of the present status of the origin-of-life problem within our current fields of science. In so doing it gives a somewhat biased—but I think realistic—view of the way the various branches of science operate in concert toward a common goal.

Vast periods of time (hundreds to thousands of millions of years) are required for a biology to develop to a stage in which it is recognizable as such. Whether the origins of life are probed from sources such as our Earth, from other planets, or from meteoritic material, only a brief instant of present time is immediately available. To study origins one must attempt both to scan and to re-create the past: to examine and date rocks, for example, in search of organic chemical and morphological evidence of predecessors. Essentially the origin-of-life problem is one of re-creating the most ancient of ancient histories. The better the ancient stage is set, the more meaningful and sound are laboratory simulations of those early events.

The hard and tedious work and the glorious insights of many researchers in this field are referred to throughout this book in most general terms. Many who have made great contributions have not been specifically mentioned; this is a limitation imposed by the brevity of the book. Others I have clearly not agreed with. I apologize to those who might be offended with the hope that they will switch (or perhaps try some more experiments) but not fight. Logical disagreement is the shortest route between minds, and proper experiments do resolve objective disagreements.

September 1978   CLAIR EDWIN FOLSOME

THE ORIGIN OF LIFE

CHAPTER ONE

A UNIVERSAL CHEMISTRY

Order is Heaven's first law.
ALEXANDER POPE

The Giant Nebula in Orion. This nebula, in the region of the "sword" of the constellation Orion, can be seen through binoculars. It is some 15 light years in diameter and contains enough gas and dust to make about 100,000 stars the size of the Sun. This nebula is one of many in our galaxy, the Milky Way, which is some 100,000 light years in diameter. The Orion nebula, 1500 light years from us, is our closest nebula. (A Lick Observatory photograph using the Mount Hamilton 120-inch reflector telescope; photo courtesy of NASA.)

Go outside and look at the stars on a clear dark night. Some are yellow or orange, others white, still others blue. Their apparent size and brightness range from those just visible to the eye to stars almost bright enough to cast shadows. A telescope shows that some of these objects are not stars at all but galaxies, huge groupings of stars. Other objects are nebulae—great incandescent clouds of gas and dust within our own galaxy.

The fantastic aspect of all this abundance and diversity is that remarkably few scientific principles suffice to explain the material composition, distances, ages, birth, and development of stars and their planets. One such principle states that the chemical elements are of universal occurrence. Reactions within and among the elements are a fundamental property of their structure, concentration, and environment. The chemistry of Earth and its Sun obeys the same laws as chemistry elsewhere throughout the Universe.

At first it might seem a prodigious task to determine the chemical composition of far distant stars—or, for that matter, of our Sun. The only contact between these bodies and the investigator is their light. But this contact has proven sufficient. A spectroscope can separate the various frequencies of light emitted by any star; the spectral lines indicate which chemical elements are present and the abundance of each.

In its most basic form, a spectroscope consists of a prism or a diffraction grating and a narrow slit interposed between the image of a star and a strip of film or some other detector. Unresolved, "white" light is a form of visible electromagnetic radiation from the star; it consists of a bundle of discrete photons. The energy of any given photon is proportionate to its frequency. Figure 1-1 depicts the relationship between energy and frequency and locates the visible spectrum within the entire electromagnetic spectrum.

Figure 1-2 shows how a diffraction slit spectroscope resolves white light, which consists of photons of many frequencies, into a spectrum of images of the slit. Each image represents a narrow range of photon frequencies.

Let us consider as a light source—a white-hot solid body. In this instance the spectrum would be smooth and continuous. Electrons of atoms within the body are boosted to many different higher orbits by the heat; they then fall again to lower orbits. The heat energy that moves an electron to a higher energy level reappears as a photon (light energy) when the electron returns to a lower energy level. The white-hot body emits photons at essentially all frequencies within the visible range.

On the other hand, if a gaseous form of any single element is heated, a "fingerprint set" of bright emission lines unique to that element appears upon an otherwise black background. Atoms of each element have a unique number of electrons and protons. When atoms of the element are heated, the electrons are excited to various energy levels about the nucleus. As excited electrons return to lower energy levels, they release this "excitation energy" by emitting photons with a set of frequencies characteristic of that element.

In the nineteenth century, Robert Bunsen and Gustav Kirchhoff used "eyeball" emission spectroscopy when they developed simple flame tests for the metals. When they burned a salt of a metal in a flame, they observed a characteristic visible

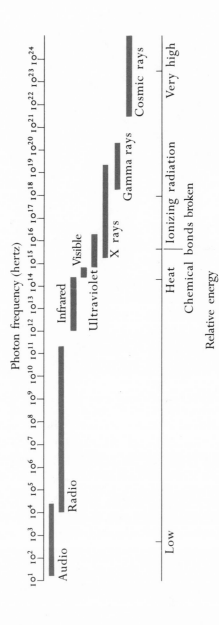

FIGURE 1-1

The electromagnetic spectrum. The spectrum ranges from below the audio frequencies to higher than cosmic rays. Of particular interest to us is the zone from the far ultraviolet to and including the visible: most of the solar energy impinging upon the Earth falls in this range.

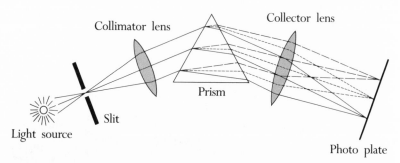

FIGURE 1-2

Diagram of a simple spectroscope. Light rays from a source enter a narrow slit and are collimated (made parallel) by a lens. Next they are beamed onto a prism (or a diffraction grating), which separates rays of different frequencies. A second lens collects the rays and beams them onto a detector. The narrower the slit, the more effective the separation of rays into frequency bands.

color for the particular metal. What they were seeing was the dominant visible color of the emission spectrum for that metal. For example, a lump of sodium chloride held in a flame appears yellow, nickel chloride appears green, and cadmium chloride appears violet-purple.

Another kind of spectrum appears as a set of dark lines on a light background. This is an absorption spectrum; it occurs when white light from an incandescent solid body passes through a gas of any element. The white light source emits photons of all frequencies. The dark lines correspond to photons that were absorbed by the low-energy-level electrons of the gas because their specific frequency bands were just right. This kind of spectrum is thus a subtractive one. For any element, emission and absorption spectral lines occur at the same sets of frequencies.

Spectroscopy is a primary point of contact between astronomy and chemistry. Analysis of light from the stars yields a

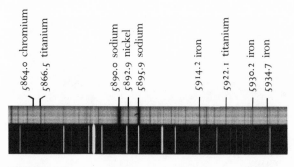

FIGURE 1-3

A portion of the yellow part of the spectrum of a K 1 giant star (top), with a comparison spectrum of neon below, from which the scale of wavelengths can be established. A number of absorption lines are marked with the wavelength and the elements that are responsible for them. (Lick Observatory spectrum, courtesy of H. Spinrad.)

wealth of chemical data. Figure 1-3 shows a typical stellar spectrum. Analysis of such spectra yields not only the identity of chemical elements but even more information as well. For example, the temperature of the source can be measured by comparing the relative intensities of different emission lines of the same element. The relative amounts of each element can be obtained by measuring the relative intensities of the major spectral lines for each element. Furthermore, the velocity of the object relative to the observer can be determined from the spectrum.

More than 2 million spectra of some 15,000 stars and of our Sun have been recorded since the late nineteenth century; from these the chemical composition of the stars and the Sun has been deduced. Perhaps equally vital is the comforting conclusion that the same physical laws and the same kinds of elements prevail throughout the Universe.

Dark-line absorption bands occur in stellar spectra when light from a distant star passes through a cloud of interstellar gas en route to the observer. Analysis of these absorption bands enables us to identify which elements and molecules are present in interstellar space.

Nebulae appear through a low-power telescope as faint stars, but they are actually vast clouds of interstellar gas and dust interspersed with stars. Radiation from these stars excites the nebular gas and dust causing it to glow. Thus the spectra of these luminous gas clouds are emission spectra.

The millions of spectra recorded from the Sun, stars, interstellar clouds, and nebulae can all be interpreted by a surprisingly short set of statements.

First, all these objects—which comprise almost all the matter of the Universe—have essentially the same chemical composition. Table 1-1 shows the relative abundance in the Universe of some of the more common elements.

TABLE 1-1

Cosmic abundance of some of the more common elements.

ATOM	RELATIVE COSMIC ABUNDANCE, ATOMS
Hydrogen	10,000,000.
Helium	1,400,000.
Lithium	0.003
Carbon	3,000.
Nitrogen	910.
Oxygen	6,800.
Neon	2,800.
Sodium	17.
Magnesium	290.
Aluminum	19.
Phosphorus	3.
Potassium	0.8
Argon	42.
Calcium	17.
Iron	80.

Second, almost all the matter of the Universe is present as the two simplest elements, hydrogen and helium. Most of the balance appears as the less complex members of the periodic table: carbon, nitrogen, oxygen, neon, magnesium, silicon, sulfur, fluorine, aluminum, chlorine, argon, calcium, and iron.

Third, the spectra of all celestial objects show a Doppler red shift: the frequencies of all photons are uniformly decreased (moved to the red side of the spectrum—Figure 1-4). More distant galaxies and stars manifest a more pronounced red shift, while closer stars show less of an effect. This means that, relative to an observer at any point in space, other stars are receding. Analysis of the magnitudes of red shifts leads us to conclude that the Universe is expanding, since all objects are becoming more distant from one another.

But why is the Universe expanding? Why does the ordering of the chemical elements and their abundances show such an apparent simplicity? Such questions are not within the scope of

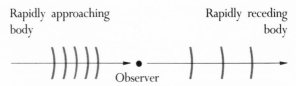

FIGURE 1-4

The Doppler effect. As a body rapidly approaches an observer, the frequencies of all its electromagnetic radiations relative to the observer are increased in direct proportion to its velocity. As a body rapidly recedes, all its electromagnetic radiation relative to the observer decreases in proportion to its velocity. Approaching bodies show a blue shift (increased frequencies) of the visible spectrum, and receding bodies manifest a red shift (decreased frequencies).

this book; answers are best sought in books about universal cosmology and theories of the origin of the Universe. Our intent here is to develop a local cosmology.

It is possible to answer the question of elemental abundances and types in a way that is consistent with the findings of nuclear physics and astronomy. However, the accuracy of this explanation rests ultimately upon a universal cosmology which has not yet been fully developed.

Hydrogen is the most abundant element, and the simplest. It consists of one proton, and one electron. If we assume that the very first primordial matter of the Universe was only hydrogen, we can account not only for all other elements, but also for their present abundances.

Given a primeval Universe of pure hydrogen, stars will form (a model of this process will be discussed in Chapter Two). Stars are essentially vast, gravitationally compacted accretions of matter, and as matter accretes, its temperature rises until nuclear reactions begin.

The most fundamental stellar nuclear reaction is hydrogen fusion. This reaction transforms hydrogen into helium and energy (Figure 1-5). The mass of the helium nucleus, which consists of two protons and two neutrons, has been accurately measured to be 4.0026 atomic mass units (amu). At temperatures and pressures high enough to trigger hydrogen fusion, four hydrogen atoms are squeezed into one helium atom. But the mass of one hydrogen atom is 1.0079 amu, and thus four hydrogens weigh 4.0316 amu. The difference between four hydrogens and one helium is 0.029 amu: a small number but one that makes the Universe go. In accord with the law of conservation of mass and energy, this mass difference reappears as energy. In Einstein's equation, energy equals mass multiplied by the square of the velocity of light. The transmutation of hydrogen to helium results in a small, 0.7% mass loss for every new atom of helium—and in the liberation of enormous amounts of energy.

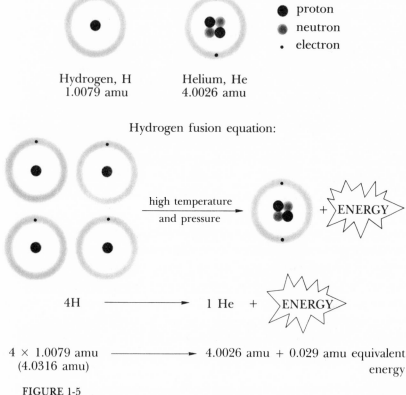

Hydrogen, H
1.0079 amu

Helium, He
4.0026 amu

proton
neutron
electron

Hydrogen fusion equation:

high temperature
and pressure

+ ENERGY

4H ⟶ 1 He + ENERGY

4 × 1.0079 amu ⟶ 4.0026 amu + 0.029 amu equivalent
(4.0316 amu) energy

FIGURE 1-5

The fusion of hydrogen to yield helium. One molecule of helium is lighter than four molecules of hydrogen; the mass deficit reappears as energy.

The occurrence of hydrogen fusion reactions in stars explains why helium, a primary product of the fusion, is observed as the second most abundant of the chemical elements. Other stellar nuclear reactions, involving helium as an ingredient, generate the higher elements or form unstable radioactive elements that decay to yield other higher elements.

Exactingly precise nuclear physics experiments have iden-
tified the most likely reactants and reaction rates in stellar
chemistry. The most fascinating aspect of this field of study is
that theoretical calculations based on these experiments can
predict the entire known array of elements, with relative abundances similar to those actually observed for the Universe.

The first all-hydrogen stars die—some become supernovae, exploding and scattering their matter back into space.
Interstellar space, which at first contained only hydrogen, now
consists of this gas plus the more complex elements, the debris
of nuclear transformations in the first stars.

The age of the Universe can be calculated using the known
abundance of the elements and known stellar nuclear reaction
rates. This age is 10,000 to 13,000 million years (m yr). Another
independent way of calculating the age of the Universe is based
on red-shift recessional velocities. If the Universe is assumed to
expand at a constant rate, this method also gives an age of 13,000
m yr.

CHAPTER TWO

AGES OF STARS AND PLANETS: THE PERIODIC TABLE

The obvious is that which is never seen until someone expresses it simply.

KAHLIL GIBRAN, "Sand and Foam"

1970 solar eclipse. A view of the Sun during the 1970 solar eclipse. The Sun, an average star, is about 300,000 times more massive than Earth. It consists mostly of gravitationally compacted hydrogen gas, and has central temperatures as high as 15 million degrees as a result of hydrogen fusion reactions. The Sun contains enough hydrogen to maintain its steady energy output for at least another 5000 million years.

*I*n principle the age (or lifetime) of a star can be determined, since stars radiate energy into space at measurable rates. Their mass energy comes from nuclear reactions, and, although the mass of a star is great, it is not infinite.

As a first approximation, consider a star of mass *m,* consisting originally of pure hydrogen. The bulk of its released energy, *E,* would then be:

$$E = (0.007)mc^2$$

where 0.007 is the mass deficiency that reappears as energy during hydrogen fusion and *c* is the velocity of light.

For a star the size of our Sun, $E = 10^{52}$ ergs. Present solar radiation is observed to be 10^{41} ergs per year. If our Sun were to emit energy at this constant rate, its life would be

$$\frac{10^{52}}{10^{41}} = 100,000 \text{ m yr}$$

(1 m yr = 1 million years).

This approximation gives far too great a lifetime, however, because the equilibrium of a star changes as hydrogen fusion continues and the star ages. Stars are massive gaseous bodies in which inward gravitational pressures are balanced by outward hydrostatic pressures. If the temperature decreases, the outward pressure decreases and the star contracts. Conversely, if the temperature increases, pressure increases and the star expands. Hydrogen fusion is temperature-dependent: its rate increases as the temperature rises.

Large stars have higher temperatures than small stars and use up hydrogen faster. Our first approximation must thus be modified to account both for changes in stellar equilibrium as the star evolves and for the mass-temperature effect.

One way to obtain a perspective on stellar evolution and to derive stellar ages is to survey a large random sample of stars. The distance, apparent brightness, and color of each star are measured. Distance can be measured by parallax for close stars, by the red-shift recessional velocity for distant stars, or by other means.

If we know the apparent brightness and distance, we can calculate the absolute magnitude, since apparent brightness equals the absolute magnitude divided by the square of the distance. The absolute magnitude is a function of the rate of energy release from a star, independent of its distance from the observer.

The color index of a star is a measure of its temperature: blue corresponds to a very hot star, white means hot, and red is relatively cool.

Figure 2-1 is a graph of absolute magnitude against color index for a large number of stars. Since it includes stars of all ages and all sizes, this classical Hertzsprung–Russell diagram depicts the "average" star at a variety of times.

Most stars are located on the straight-line portion of the diagram; they are transmuting their hydrogen with only gradual changes in equilibrium. On this portion of the graph, called the main sequence, the more massive stars have higher temperatures, faster rates of hydrogen fusion, and shorter lifetimes. Stars

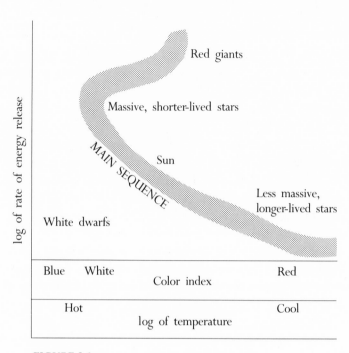

FIGURE 2-1

The Hertsprung—Russell diagram showing absolute magnitude versus color temperature for a large number of stars within our galaxy. Many stars were sampled for this diagram; any single point represents one star. Since stars were sampled at random, the graph indicates the average history of a star.

less massive than the Sun are at lower temperatures; hydrogen fusion occurs at slower rates and lifetimes are longer. When any star on the main sequence has consumed about 10% of its initial hydrogen, its temperature decreases and it expands. The red giants are thought to be old main-sequence stars of all sizes.

These factors must all be taken into account in making accurate determinations of the ages of stars. Modified calculations show that no star of our galaxy is older than some 11,000 m yr. Some small stars are this old; many larger stars are much younger. The most massive stars spend no more than 1 m yr on the main sequence. Our Sun and stars of its size will spend about 10,000 m yr there before entering the red giant stage. This calculated age is only 10% of our first approximation of 100,000 m yr in which all the hydrogen was assumed to be used. The second approximation could be (and has been) refined further, but those corrections have little bearing on our views on stellar evolution as it pertains to the origin-of-life question.

Once the maximum lifetimes of stars and the Sun are determined, the next problem is to measure the ages of our planet and similar bodies. But before this question can be answered it must be phrased more carefully. It is possible to determine how long a particular crystallized mineral has remained in that state, or how long a certain meteorite has been subjected to cosmic ray bombardment, or even (on a far less majestic scale) how old is a given piece of wood. In fact, these how-long-since-the-event determinations are what give "ages." The basis of these determinations lies in the chemistry of the atom.

Each atom of every chemical element consists of a central nucleus, made up of protons and neutrons, and an outer cloud of negatively charged, much less massive electrons. One of the triumphs of chemistry has been the elucidation of the physical basis for the empirical fact that the elements can be arranged in a periodic table, in which they are ordered in sequence according to the number of protons per nucleus (Figure 2-2). When thus arrayed, the elements show regularly recurring similarities in chemical properties.

Hydrogen, the simplest element, consists of a single proton associated with one electron. But the number of neutrons present in the hydrogen nucleus is not fixed—it can vary. These different forms of hydrogen—called *isotopes*—vary in mass from 1 amu (atomic mass unit) to 3 amu (see Figure 2-3). The various

Periodic Table

IA	IIA	IIIB	IVB	VB	VIB	VIIB	VIIIB			IB	IIB	IIIA	IVA	VA	VIA	VIIA	O
1	1																2
1 H 1.008																	He 4.003
2	3 Li 6.941	4 Be 9.012											5 B 10.81	6 C 12.01	7 N 14.01	8 O 16.00	9 F 19.00	10 Ne 20.18
3	11 Na 22.99	12 Mg 24.31											13 Al 26.98	14 Si 28.09	15 P 30.97	16 S 32.06	17 Cl 35.45	18 A 39.95
4	19 K 39.10	20 Ca 40.08	21 Sc 44.96	22 Ti 47.90	23 V 50.94	24 Cr 52.00	25 Mn 54.94	26 Fe 55.85 · 27 Co 58.93 · 28 Ni 58.71			29 Cu 63.55	30 Zn 65.37	31 Ga 69.72	32 Ge 72.50	33 As 74.92	34 Se 78.96	35 Br 79.90	36 Kr 83.80
5	37 Rb 85.47	38 Sr 87.62	39 Y 88.91	40 Zr 91.22	41 Nb 92.91	42 Mo 95.94	43 Tc 98.91	44 Ru 101.1 · 45 Rh 102.9 · 46 Pd 106.4			47 Ag 107.9	48 Cd 112.4	49 In 114.8	50 Sn 118.7	51 Sb 121.8	52 Te 127.6	53 I 126.9	54 Xe 131.1
6	55 Cs 132.9	56 Ba 137.3	* 71 Lu 175.0	72 Hf 178.5	73 Ta 180.9	74 W 183.9	75 Re 186.2	76 Os 190.2 · 77 Ir 192.2 · 78 Pt 195.1			79 Au 197.0	80 Hg 200.6	81 Tl 204.4	82 Pb 207.2	83 Bi 209.0	84 Po (210)	85 At (210)	86 Rn (222)
7	87 Fr (223)	88 Ra 226.0	† 103 Lw (257)	—	—													

* Lanthanides

57 La 138.9	58 Ce 140.1	59 Pr 140.9	60 Nd 144.2	61 Pm (147)	62 Sm 150.4	63 Eu 152.0	64 Gd 157.3	65 Tb 158.9	66 Dy 162.5	67 Ho 164.9	68 Er 167.3	69 Tm 168.9	70 Yb 173.0

† Actinides

89 Ac (227)	90 Th 232.0	91 Pa 231.0	92 U 238.0	93 Np 237.0	94 Pu (242)	95 Am (243)	96 Cm (248)	97 Bk (247)	98 Cf (251)	99 Es (254)	100 Fm (253)	101 Md (256)	102 No (254)

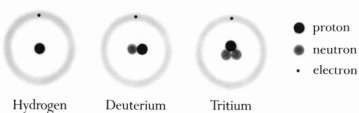

Hydrogen
1 amu
Stable

Deuterium
2 amu
Stable

Tritium
3 amu
Radioactive

● proton

◉ neutron

· electron

FIGURE 2-3

The isotopes of hydrogen. All three hydrogen isotopes possess a single-orbiting electron, but they differ in their nuclear mass (and in their stability). Stable hydrogen and deuterium possess, respectively, one proton and one proton plus one neutron; unstable (radioactive) tritium has one proton and two neutrons in its nucleus.

hydrogen isotopes can be separated and the amount of each measured.

The various isotopes of an element correspond to different numbers of neutrons present within the nucleus. Although the chemistry of the isotopes of any given element is essentially identical, two major points of difference form the basis for age-dating schemes.

First, in the ultramicro world of atoms and molecules, chemical reactions are related to the speed of the atoms or molecules. Given similar accelerations, heavier particles attain less velocity than lighter ones. As a result, lighter isotopes enter more easily into chemical reactions that proceed quickly in one

FIGURE 2-2

The periodic table of the elements. All the elements can be placed in this table by virtue of the number of protons within their nuclei and the arrangement of the electrons in their outer shells. Predictions about the chemical behavior of any element can be made on the basis of its position within the periodic table.

direction. Hence, in biological systems, the nitrogen isotope N^{14} reacts more easily than the heavier isotope N^{15}, and so biological systems tend to concentrate N^{14} more than N^{15}. Similarly, C^{12} is enriched relative to C^{13} and O^{16} is enriched relative to O^{18}. This selective enrichment for light isotopes by biological systems is easily measurable.

The second property of interest is the stability of isotopes. Physical theory cannot yet explain why, but the fact is that certain combinations of protons and neutrons are stable, whereas others are unstable and undergo radioactive decay. Both hydrogen (H^1) and deuterium (H^2) are stable, but tritium (H^3) decays, violently releasing a high-speed electron called a beta particle; in the process, H^3 becomes helium-3 (He^3). Carbon isotopes C^{12} and C^{13} are stable, but C^{14} decays, releasing a beta particle to become stable nitrogen-14 (N^{14}).

Likewise, the potassium isotopes K^{39} and K^{41} are stable, but K^{40} decays by giving off a beta particle to become calcium-40. Potassium-40 also has another mode of decay (called electron capture) in which the nucleus captures one of its orbiting electrons, which combines with a nuclear proton to become a neutron. The product in this case is argon-40.

Many of the more complex elements, with larger numbers of protons and neutrons, decay into products which are themselves radioactive. Such series decay is typified by the behavior of uranium-235, uranium-238, and thorium-232. Frequently these decay processes emit energetic alpha particles (helium nuclei, consisting of two neutrons combined with two protons).

Radioactive isotope decay processes can be grouped by the mode of decay (Figure 2-4). A nucleus that emits an electron as a beta particle does not simply cast off an orbiting electron, but instead manufactures it in the nucleus by the decay of a neutron into a proton and an electron. This process, called beta decay, increases the atomic number of the atom by one, transforming it into a different element. Alpha decay lowers the atomic number

A = atomic
   number

Z = atomic
   weight

Alpha decay example

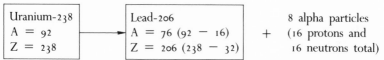

Uranium-238		Lead-206			8 alpha particles
A = 92	→	A = 76 (92 − 16)	+		(16 protons and
Z = 238		Z = 206 (238 − 32)			16 neutrons total)

Beta decay example

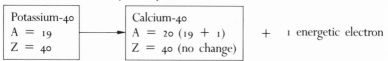

Potassium-40		Calcium-40		
A = 19	→	A = 20 (19 + 1)	+	1 energetic electron
Z = 40		Z = 40 (no change)		

Electron capture example

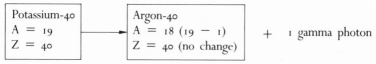

Potassium-40		Argon-40		
A = 19	→	A = 18 (19 − 1)	+	1 gamma photon
Z = 40		Z = 40 (no change)		

FIGURE 2-4

Mechanisms of radioisotope decay. Radioactive isotopes decay by three basic mechanisms: alpha particle emission, in which the nucleus simultaneously loses two protons and two neutrons (an alpha particle); beta emission, by which one energetic electron is lost; and electron capture, by which one energetic (gamma) photon is lost.

by two (two protons lost) and the mass by four (two protons plus two neutrons) for every alpha particle ejected. Electron capture (or K-capture) is the inverse process of beta emission, since a proton of the nucleus captures an inner electron of the K orbit and thus becomes a neutron. In K-capture the atomic number is reduced by one and the mass remains constant.

    The rate at which a radioactive isotope of an element decays is a unique constant for that isotope; its value is not disturbed by any natural environmental change. Decay rates are commonly expressed as half-lives; an isotope's half-life is the time required for half of a given amount of it to decay. Although the exact moment of decay for any single atom is impossible to predict, half-lives can be determined with considerable precision because the process obeys statistical principles. Large numbers of atoms are present even in minute samples of matter. One gram of carbon, for example, contains some 5×10^{22} atoms.

Table 2-1 shows half-lives and decay constants for some of the elements most commonly used in age dating. The essence of most dating schemes is to measure the amount of parent and/or daughter isotopes with exquisite accuracy. The ratio of the isotopes, placed in the proper formula, then yields an estimate of the age of a substance. This age represents the time that the mineral has existed in its crystalline form. Several examples are given below.

The carbon-14 clock, developed by Willard Libby, is most useful in archaeology and more recent events, but its principles illustrate those of dating methods in general. The earth is being bombarded by high-energy cosmic-ray particles at a constant rate. These high-velocity particles, mostly protons, interact with molecules of the atmosphere to produce neutrons with sufficient energy to be absorbed by nitrogen-14 nuclei. Every nitrogen-14 atom thus affected is transformed into carbon-14, a radioactive isotope whose half-life is 5730 years.

Carbon-14 is formed and lost at fixed rates. With highly specialized Geiger counters and accurate chemical determinations for total carbon, the abundance of C^{14} relative to all isotopes of carbon can be measured. The carbon pool of the Earth—carbon dioxide in the air and oceans, carbonate minerals in rocks, and living organisms—contains an equilibrium amount of carbon-14 which amounts to 1.2×10^{-12} gram of carbon-14 per gram of total carbon.

When an organism dies, all its carbon is removed temporarily from the carbon cycle of the world. Its carbon-14

TABLE 2-1

Half-lives and other data for some elements useful in age determination.

FROM	TO	DECAY PROCESS	HALF-LIFE[a]	DECAY CONSTANT[b]
Carbon-14	Nitrogen-14	Beta emission	5.7×10^3	1.216×10^{-4}
Potassium-40	Argon-40	Electron capture	1.3×10^9	0.585×10^{-10}
	Calcium-40	Beta emission	(combined)	4.720×10^{-10}
Rubidium-87	Strontium-87	Beta emission	4.7×10^{10}	1.474×10^{-11}
Thorium-232	Lead-208 and 6 alpha particles	Alpha emission (series decay)	1.39×10^{10}	4.99×10^{-11}
Uranium-235	Lead-207 and 7 alpha particles	Alpha emission (series decay)	7.13×10^8	9.72×10^{-10}
Uranium-238	Lead-206 and 8 alpha particles	Alpha emission (series decay)	4.51×10^9	1.537×10^{-10}

[a] Half-lives (in years) are expressed in exponential notation; for example, 10^3 = 1000 years, 10^9 = 1,000,000,000 years.

[b] For some uses, decay constants rather than half-lives are employed. The decay constant d is related to the half life by:

$$\text{half-life} = \frac{(\ln 2)}{d} = \frac{(0.693)}{d}$$

content becomes depleted, since no new carbon can enter the system. For a given sample, measurements of carbon-14 and total carbon can be combined with data for the carbon half-life (a constant) and the equilibrium amount of carbon-14 (another constant) to give the age of the sample from the time it was no longer open to the carbon cycle. In practice the accuracy and scope of measurements are limited. The older the sample, the less carbon-14 activity remains. The limit for the carbon system is about ten half-lives, or 57,300 years, at which time only about 0.001 of the equilibrium value (0.016 disintegrations per second per gram) remains. This length of time is but a recent instant in the chronology of the Earth. To date ancient rocks, we must find other methods. The principle of using radioactive isotopes or their products remains useful, but isotopes with long half-lives are required.

Consider a uranium-bearing rock. Assume that all parent isotopes entered the rock at the time of crystallization and that neither parent nor daughter isotopes have been lost from that time to the present. (In practice, departures from these assumptions occur and can be dealt with.) Uranium/thorium-bearing rocks contain both isotopes of uranium (U^{238}, U^{235}) and thorium-232, as well as the four isotopes of lead (Pb^{204}, Pb^{206}, Pb^{207}, Pb^{208}). Lead-204 is not produced by any known radioactive source but appears to be primordial. The other three lead isotopes are derived from decay of the uranium/thorium isotopes.

Table 2-2 shows the parents and half-lives of various lead isotopes. In the ideal case the ratios of lead to uranium (or thorium) will all yield the same age to corroborate age determinations. The older the rock, the less uranium and the more lead is present. The Pb^{207}/Pb^{206} ratio itself can be used as a clock, since uranium-235 has a much shorter half-life than uranium-238.

A large number of age determinations have been carried out in this fashion, and in general the results are internally consistent. The key to success in this method is the analysis of uranium-bearing minerals which, at the time of crystallization, contained little or no lead.

Since the ratio of lead-207 to lead-206 can itself serve as a clock, samples from the Earth's crust, deep ocean sediments, and elsewhere have been surveyed to obtain the average value for this ratio, and to arrive at an estimate of the average age of the Earth. This number is 4500 m yr. An essentially similar approach, using stony and iron meteorites, also yields an age of 4500 m yr.

TABLE 2-2

Sources of lead isotopes found in uranium/thorium-bearing minerals. The percentages of lead isotopes are those in the Canyon Diablo meteorite.

PRECURSORS	HALF-LIFE	LEAD ISOTOPES (ABUNDANCE)
No known parent	—	Lead-204 (2.0%)
Uranium-238 (intermediate abundance)	4.5×10^9 yr	Lead-206 (18.8%)
Uranium-235 (least relative abundance)	7.13×10^8 yr	Lead-207 (20.6%)
Thorium-232 (greatest relative abundance)	1.39×10^{10} yr	Lead-208 (58.6%)

These figures represent the time since the oldest rocks found on the Earth became crystallized, or the time since meteorites formed. The age of more recently crystallized rocks (from 4500 m yr ago to about 500 m yr ago) can be determined with equivalent accuracy (about 4% or better). Rocks formed during the period from 500 m yr ago to the present are mainly sedimentary; these present special problems in dating by the uranium—lead method. However, this particular time span, recent in geological perspective, has been given a time scale in terms of sedimentary rock formations and biological fossil sequences. Table 2-3 summarizes ages, age-dating methods, and major features of the entire time scale.

The lead—uranium method of age determination can be corroborated externally as well as internally. Other minerals that are deficient in uranium but that contain potassium-40 or rubidium-87 can be used for this purpose. For example,

TABLE 2-3

Time scale for the Earth.

ERA	AGE (M YR)	EFFECTIVE DATING METHOD	MAJOR TERRESTRIAL FEATURES
Recent	0—1	Carbon-14	Abundant, complex fossil record
Cenozoic	1—62	Sedimentary strata, fossil sequences	
Mesozoic	62—230		
Paleozoic	230—570	Radiometric ages	
Proterozoic	570—2600	Uranium/lead, potassium/argon, rubidium/strontium	Microfossils, first free oxygen
Archean	2600—3600		Protobionts
Pre-Archean	3600—4500		Chemical evolution

potassium-40 decays slowly to both calcium-40 and argon-40. The argon-40 can be released from the matrix of the rock and its abundance measured. Measurement of the amounts of potassium-40 and argon-40 in a mineral indicates the age of crystallization of the mineral.

The dating methods just discussed have yielded a time scale for the Earth, from its cooling 4500 m yr ago up to the present. Within this time scale, the task is now to determine what conditions on the primitive Earth really were. What kind of atmosphere did it have? What was the temperature? The pressure? When did the oceans form? How did the Earth itself form?

CHAPTER THREE

THE FORMATION OF PLANETARY SYSTEMS

We are men, and our lot is to learn
and be hurled into inconceivably new worlds.

G. B. SHAW, *Man and Superman*

Jupiter and its moons. (Photo courtesy of NASA.)

*C*ertain fundamental observations made in the preceding chapters permit us to phrase and deal with origin questions more precisely.

First, there is a universal chemistry, manifest to us through analysis of starlight. This implies that the same elements and their reactions obey the same physical laws for any star and in any galaxy. At present we have no reason not to believe that all physical phenomena—electromagnetic radiation, radioisotope decay, gravity, and so on—operate uniformly throughout the Universe.

Second, the life spans of stars, including our Sun, are calculable. Generally, large stars have short lives and small stars have long lives. A G-type star, such as the Sun, has a main-sequence life of about 10,000 m yr.

These two generalizations are based upon observation and reason, but to use them profitably in posing origin questions, we must also make use of a philosophical "principle." This is most simply stated as, "No observed process can ever be explained by accidental occurrences." Were we to admit the possibility of pure and simple accident rather than attempting to find an underlying reason for a process, we would be unable to formulate laws and generalizations about the process or test it by experiment.

Accidental chance and statistical chance are two quite

different notions. As used probabilistically, the idea of chance gives us a way of learning about events that may have any one of several outcomes. For example, with the present state of our knowledge we cannot predict when any one specific radioactive atom will decay. But for a large number of such atoms, we do know quite precisely what the average number of atoms decaying in a given period will be.

An example of the improper use of chance would be to use it to explain the accidental occurrence of a particular event in a historical context. Consider the fact that amino acids occur in two forms—denoted L and D—which are mirror images of each other (see Chapter Twelve for more detail). However, only the L forms are found in all proteins of all living things. Mirror image molecules were long considered to have identical physical properties. Hence one is sorely tempted to "explain" the fact that our biology employs L and not D amino acids by postulating a historical accident during evolution. But this approach generates for us a uniqueness we may not deserve. Also, if we adopt this philosophy of accidental chance we no longer have recourse to experiment. The alternative is to assume that laws must exist, of which we are not yet fully aware, that favor the use of the L amino acids.

The philosophical stance just adopted becomes relevant to the origin-of-life problem when we consider models and theories for the origin of planetary systems. Basically, two types of theories have been proposed.

Catastrophic rare or near-unique events represent one approach. In these models our Sun undergoes a near miss or a grazing collision with another star. Alternatively, the Sun passes, quite by accident, through a dense cloud of interstellar gas and dust. In any event, the outcome is a unique set of planets of which Earth is the most unusual. This model type is a typical accidental-chance "explanation," and we must rule it out of further consideration in our search for general laws.

The second class of models for the origin of planetary

systems states that planets are a common byproduct of star formation. Considerable evidence supports this view, which was first proposed by Immanuel Kant in the eighteenth century and later modified by Gerard Kuiper, H. Alfvén, and A. G. W. Cameron.

Young stars are found within nebulae, light-year-size patches of relatively concentrated interstellar dust and gas. Nebulae occur throughout our galaxy, and within these vast clouds of matter stars and their associated planetary systems are believed to be forming.

Spectroscopy shows that interstellar matter consists of hydrogen, helium, and neon as gases, and micron-size grains of dust consisting of metallic and other elements. Since the temperature is very low (10 to 20°K), all material except the gases just mentioned occurs in frozen form on the dust grains. The heavier elements and some of the hydrogen originated from preceding eras of stars, some of which exploded as supernovae, returning their remaining hydrogen and their hydrogen-transmuted heavier elements into space.

Interstellar space has an average gas density of only 0.1 H atom per cubic centimeter. Nebulae can have gas densities of 1000 H atoms per cubic centimeter, a 10,000-fold increment in concentration. (One cubic centimeter of air contains some 2.7×10^{19} molecules.) Although by our standards the nebulae themselves would constitute a "vacuum," their vast size, measured in light years, supplies more than enough matter to make thousands of stars and planetary systems the size of ours.

When a vast nebular cloud becomes large enough, by slowly accreting interstellar gas and dust by gravitation, it becomes unstable; the near balance between pressure and gravitational forces is disrupted. Gravitational forces predominate, and the cloud contracts. During the early phases of contraction, heat formed from conversion of gravitational potential energy to radiant energy leaves the cloud easily, since the relative density of matter and gas is low. As the density of matter increases, an important new change begins.

Because of gravitational and other fluctuations, the large cloud breaks up into smaller clouds which in turn fragment, ultimately to masses and sizes several times greater than those of our solar system (Figure 3-1). Such a cloud is termed a protostar. Of course, some protostars will be more massive than our system; these will form larger, hotter stars that will evolve faster. Others, somewhat less massive, will form smaller, cooler stars that will evolve more slowly. The sizes of protostars range from

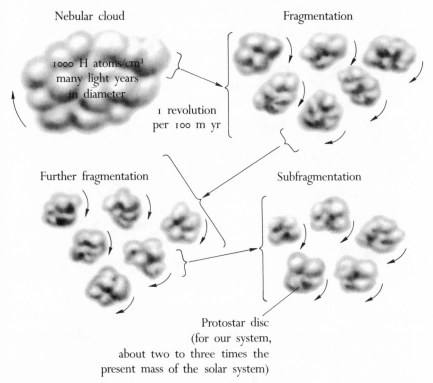

Nebular cloud

1000 H atoms/cm³
many light years
in diameter

1 revolution
per 100 m yr

Fragmentation

Further fragmentation

Subfragmentation

Protostar disc
(for our system,
about two to three times the
present mass of the solar system)

FIGURE 3-1

Star and protoplanetary system formation by fragmentation of nebular clouds.

an upper limit above which further fragmentation would occur, to a lower limit set by the minimum amount of mass required to sustain nuclear reactions.

At first, gravitational potential energy converted to heat (radiant energy) during gravitational collapse simply radiates away. But as the concentration of matter increases, more of the radiation is absorbed, and the temperature consequently rises. Thus volatile compounds originally frozen on dust grains evaporate. Gases such as ammonia (NH_3), methane (CH_4), water (H_2O), and hydrogen cyanide (HCN) are now intermixed with H_2, He, and Ne. These gases absorb further radiant energy to become dissociated and ionized.

Gravitational collapse continues as long as the resultant radiant energy is dissipated in the evaporation and ionization of molecules on the dust grains. When the molecules are fully ionized, the temperature mounts rapidly until collapse almost halts because the gas pressure counterbalances the force of gravitation. The phase of *rapid* gravitational collapse thus comes to an end.

At this point in its evolution, a protostar representative of our system would resemble a disc with a large center having about the same radius as the solar system but containing at least three times its present mass. Temperatures would be about 10,000 to 20,000°K at the center and roughly 1000°K near Jupiter's orbit.

Such a protostar disc would then evolve by rearrangement and slower contraction. The protostar itself would slowly become more concentrated, more massive, and hotter, since heat can now be radiated only from its surface. Convection currents would serve to transport heat to the surface from within. From the surface of the protostar, a haze of gas and dust would extend outward for a distance equivalent to the orbit of Pluto.

During this complex series of contractions and collapses (which is thought to require at least 100,000 years, and more usually 10 m yr), angular momentum must be conserved. The whole galaxy, of which nebulae and dust clouds are a part, rotates with a period of one revolution per 100 m yr. As dust clouds contract, their angular momentum cannot change: the more they contract, the faster they spin. Conservation of angular momentum also results in a shape change of the contracting dust cloud from spherical to discoidal. It is thought that ultimately two kinds of discs might result: a flat, pancake-shaped kind that could give rise to binary star systems, and a fried-egg-shaped alternative that could form a single star and its planetary system.

The gas and dust particles in the flat edges of the spinning protostar interact with one another by turbulence and viscous drag. The result is that rotation of the outer edges speeds up and rotation in the center slows down. The protostar grows slowly by accretion as much gas and dust of the slowed inner zone falls into smaller and smaller orbits.

Observations of nebulae in which stars are now forming show the presence of a unique kind of star, called the T-Tauri type. One can see vast clouds of matter violently expelled by these beginning stars. Hence, we hypothesize that at some time during the slow contraction phase, an explosive "solar-wind evaporation" occurs, blowing much of the matter of the protostar back into interstellar space. What is left of the protostar is one-third to one-fourth of its original mass. This material is all that remains for further evolution.

As the remaining protostar matter continues its contraction, temperatures become high enough to trigger hydrogen fusion. With the greater energy supplied by this reaction, temperatures are now high enough to balance further gravitational contraction forces. The protostar has entered the main sequence of stellar evolution.

Planets form from the leftover gas and dust in the outer edges of the protostar disc. Although we tend to think of planets as massive objects, all our planets represent only 0.135% of the total mass of the solar system. Our planets, and presumably those formed from any protostar disc, are located in two major zones. The inner zone—extending in our case from Mercury to the asteroids—is the small, terrestrial planet zone. Here, during the slow contraction phase of the protostar, temperatures are so high that metals are vaporized. In contrast, the cold outer zone contains gases such as H_2, He, and Ne and dust grains covered with such frozen volatile substances as H_2O, NH_3, and CH_4. This outer, Jovian planet zone contains a great deal more material. This is because the volume of that zone is greater (Figure 3-2) and because much of the volatile material originally in the inner zone was pushed out by protostar activity. Due to the consistently lower overall temperatures of the outer reaches of the system, the relative abundance of the elements in material of the Jovian zone resembles that found in nebulae and stars.

The hot small-planet zone cools quickly, but its material represents dust grains and aggregates of mostly nonvolatile, refractory, and dense "earthy" matter, largely devoid of volatile material and gases.

This is the most likely reason why we (and probably all other single star systems) have two kinds of planets: the inner planets—small, dense "terrestrial" ones—and the outer cool, light, large planets.

Gravitational forces alone cannot account for accretion of nebular dust grains until the protoplanet reaches a critical mass. To begin with, smaller aggregates (Figure 3-3) most probably form by the clumping together of dust grains with tarry, sticky organic matter, with ice, and by ferromagnetic and electrostatic forces, all acting in concert as a kind of protoplanetary glue. By these mechanisms, aggregates continually form and increase in size. Throughout this time, fluctuations in energy emission by the protostar as it prepares to enter the main sequence cause nearby aggregates to melt. Larger aggregates also melt because

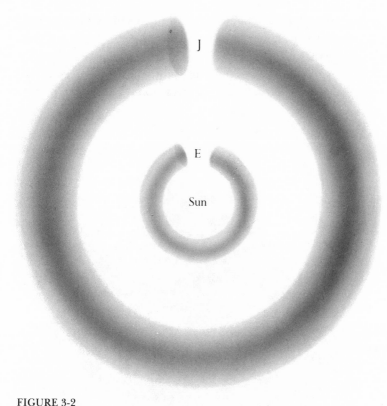

FIGURE 3-2

Why the outer planets are larger. The volume of space swept out by an inner planet (E) is much less than the volume for an outer planet (J). On this basis alone, outer planets would contain considerably more matter, assuming that the distribution of gas and dust particles is the same throughout the protostar disc.

of the internal heat derived from the decay of short-lived radioisotopes such as aluminum-26. (The short-lived elements were formed by nuclear reactions during the early, violent transition of the protostar to a star. These elements, as well as many other protostar compounds, were thrown out as the protostar entered the main sequence.)

Further melting, remelting, and aggregate building leads to the formation of planetoids having sufficient mass to accumulate large aggregates by gravitational attraction. Consonant with this

FIGURE 3-3

The Murchison meteorite. On September 28, 1969, a meteorite fell near the town of Murchison, Australia. About 10,000 meteorites strike the Earth every year, although far fewer are recovered. With the exception of lunar rocks returned by the Apollo missions, meteorites are the only source of extraterrestrial material available for us to study. Meteorites are fragments of aggregates of preplanetary material; such aggregates range in size from dust specks to objects as massive as the asteroids Icarus and Ceres. (Photo courtesy of NASA.)

scheme is the cratering seen on all the terrestrial planets and on the satellites of Earth and Mars (Figure 3-4).

FIGURE 3-4

The Moon. Dominant features are numerous meteorite impact craters of all sizes interspersed with relatively flat areas—the maria—which were once thought to be lava flows or volcanic ash but which are more probably the dust debris of impact phenomena. The craters, as old as 4600 m yr, reflect the process of planet formation by accretion of aggregates. (Photo courtesy of NASA.)

A planetoid the size of a large asteroid will cool within several million years. A planet the size of Earth will melt soon after its formation and will require a considerably longer time to cool and evolve.

Presumably, the erratic and violent T-Tauri type solar winds of the protostar blew away any remaining gas and dust that are gravitationally trapped by the forming planets. Some of these ejecta form the comets of our solar system.

This scenario of planetary system formation postulates that planets are a normal product of star formation and that two fundamentally different kinds of planets will occur: the dense, small terrestrial planets and the large, light Jovian ones.

What evidence have we that this might indeed be the case? Optical astronomy cannot visualize planets that ought to be associated with stars. It has been possible, as Peter van de Kamp has shown, to study close stars such as Barnard's star for oscillations that do exist and that can only be explained by gravitational interaction of the star with at least one planet of Jovian size. However, this method is tedious and cannot be applied to the more distant stars.

Again, spectroscopy provides a reassuring answer. If the general theory of planetary system formation is correct, the angular momentum of a star must be low and that of its planets high. Hence, if we can compare the spin rates of stars with that of our Sun, which we know has planets, then we should be able to recognize planet-bearing stars.

We can measure the spin rate of a star by making shrewd use of the Doppler effect. Figure 3-5 illustrates this. Consider a rapidly rotating star. Photons emitted from its receding side will be slowed down, while those from its advancing side will be accelerated. If the width of any stellar spectral line is measured, it will tend to be broadened more for stars with high angular momentum and less for stars with low angular momentum—those bearing planets.

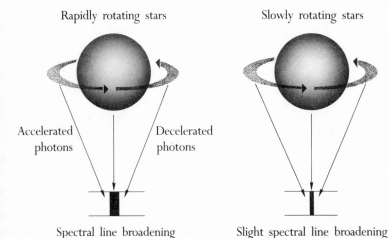

Rapidly rotating stars

Slowly rotating stars

Accelerated
photons

Decelerated
photons

Spectral line broadening

Slight spectral line broadening

FIGURE 3-5

*Rapidly rotating stars show substantial spectral line broadening. Slowly
rotating stars manifest slight spectral line broadening. Most stars the size of
the Sun show slight spectral line broadening; thus, most stars rotate slowly
and possess planets.*

Such a survey has been conducted. Stars that are massive,
hot, and quickly evolving were found to be rotating rapidly.
Those stars of spectral types F-2 to M—including our Sun and
more than 93% of all main sequence stars—are rotating slowly.
Where has all the angular momentum gone? Into planets. Our
galaxy contains some 10^{11} stars, and by this criterion almost all
should possess planetary systems.

CHAPTER FOUR

ORIGIN AND EVOLUTION OF TERRESTRIAL ATMOSPHERES

In the beginning, this world was nothing at all,
heaven was not, nor earth, nor space.
Because it was not, it bethought itself:
I will be. It emitted heat.

An Egyptian text

Surtsey eruption. This volcanic eruption near Iceland created a new island of volcanic rock and ash and provided biologists with an opportunity to study the way organisms invade an area. (Photo courtesy of NASA.)

The primeval Earth consisted of a spherical jumble of aggregate upon aggregate, heated to a largely molten state by release of gravitational potential energy, by short-lived radioactive decay, and by the variable radiation of a Sun entering upon the main sequence. There was no gaseous atmosphere to begin with; the Earth roughly 4000 m yr ago consisted largely of metallic oxides, metallic salts, metal carbonates and carbides, and metallic hydrides, as well as trapped pockets of volatiles which were the original "glue." All the past and present atmosphere of the Earth, as well as other terrestrial planets, was and is derived from volcanoes spewing forth volatile materials, which are then modified by the conditions of the environment.

Heinrich Holland has elegantly derived from theory the most likely volcanic gases that might have been emitted during various stages of our planet's evolution, but considerable difference of opinion exists concerning the immediate and long-term fate of these gases.

Many follow the original reasoning of Harold Urey and A. I. Oparin, which postulates a primeval atmosphere of methane, ammonia, and water. Originally this conclusion arose from consideration of cosmic elemental abundance data and the realization that hydrogen was relatively the most abundant element. We have seen that the terrestrial planets form mainly from nonvolatile refractory aggregates; in these, hydrogen, although present, would not be a grossly abundant element. Consider, for example, how rare other cosmically abundant volatiles such as helium, neon, krypton, xenon, and radon are on the Earth. However, hydrogen would be present in sufficient amounts that *if in equilibrium* all forms of carbon and nitrogen would be fully reduced to methane and ammonia. However, to attain chemical equilibrium requires time, and on the primitive Earth new volcanic gases were constantly being added to the atmosphere. While volcanic gases are in equilibrium at their $1200°C$ emission temperatures, this mixture is far from equilibrium at the $25°C$ temperature of the atmosphere.

Because of these considerations, some researchers have a predilection to "start" chemical evolution experiments with a first atmosphere containing these fully reduced compounds. But nature, though ordered, is of great complexity, and it seems doubtful that reactions between carbon and nitrogen (emitted from volcanoes as carbon monoxide, carbon dioxide, and nitrogen gas) and hydrogen would lead quickly to equilibrium to form only methane and ammonia. Harold Urey proposed this thought in 1953; he considered the alternative that "complex tarry intermediates" would probably exist along this path to equilibrium. In practice it seems reasonable to conceive of the early atmosphere as a complex consisting of fresh volcanic gases along with some highly reduced products and some tarry intermediates.

The Holland model proposes three pivotal stages. The first, which occurs before formation of the Earth's core, is termed the hot primordial volcanic stage. The second is called the postcore Hawaiian volcanic stage, and the last is the biological stage. Let us consider the essentials of each in turn.

During the hot primordial volcanic stage, which probably lasted from 4500 m yr to 4000 m yr ago, the core of the Earth had not formed, and the interior of the planet was undergoing rapid, massive structural rearrangements. Volcanic outgassing was the major activity and surface feature during the 500 million years of this period. Today we can construct mixtures of known Earth minerals and melt these at 1200°C, the temperature of volcanic magmas. Basic chemistry allows us to predict the likely ratios of various gases that will be emitted from the magma. The key to these predictions is the amount of oxygen present. During the precore stage of our planet's history, much metallic iron—which is now in the core—was mixed throughout. If we follow Holland and mix this iron into such other minerals as silicon and magnesium ores of current volcanic melts, we find that the pressure of oxygen during the precore stage was extremely low—$10^{-12.5}$ atmospheres.

Once we know the amount of oxygen in primordial volcanic emissions, chemical equilibrium arguments allow us to calculate the ratios of various common volcanic gases and to predict what gases might have been present in quantity at magma temperatures of 1200°C.

Consider these reversible reactions:

$$H_2 + \tfrac{1}{2}(O_2) \underset{k_1}{\overset{k}{\rightleftharpoons}} H_2O$$

$$CO + \tfrac{1}{2}(O_2) \underset{k_3}{\overset{k_2}{\rightleftharpoons}} CO_2$$

Since we have laboratory data on reaction rate constants (the k's in the above equations), if we know the amount of oxygen we can derive the ratio of the other two compounds for each of the reactions listed above, and for others involving oxygen as well.

When the oxygen pressure is $10^{-12.5}$ atmospheres, the hydrogen-to-water ratio is about 2 to 1: twice as much hydrogen as water vapor is given off. Similarly, the ratio of carbon monoxide to carbon dioxide is about 5 to 1. Methane is found to

TABLE 4-1

Early volcanic outgassing atmosphere of the primitive Earth.

MAJOR GASES	MINOR GASES	TRACE GASES
Hydrogen	Carbon dioxide	Methane
Water vapor	Sulfur	Sulfur dioxide
Nitrogen		
Carbon monoxide		
Hydrogen sulfide		

be a minor constituent: the ratio of methane to carbon dioxide is 0.0019 to 1. Table 4-1 gives the most probable composition of the gases ejected from early volcanoes.

It is difficult to guess at the atmospheric pressure of this gas mixture, but a likely range is from one-half to twice the present atmospheric pressure.

The average surface temperature of the primitive Earth was close to its present value of $25°C$, and the total outgassing atmospheric pressure of volcanic origin was fairly high. The water, ejected as vapor, condensed as massive and nearly continuous rainstorms. Shallow pools and lakes formed, and the rain washed from the sky some of the particulate volcanic ash that formed a persistent red haze.

As depicted, this atmosphere was poisonous, yet it was necessary to the ultimate occurrence of life. A human could exist on the primeval Earth only by breathing from a scuba tank and wearing protective garments against the Sun's ultraviolet rays. Water in the forming pools and lakes would be muddy but fresh, not salty, since it had no time to dissolve salts or weather rocks significantly, and since it contains much fine ash from the sky.

Given this chemical setting, let us now consider the energy sources acting on the primitive Earth. The Sun was shining then as now. But then it shone through an entirely different atmospheric window. Today our atmosphere contains oxygen, replenished continuously by photosynthesizing plants. In the upper atmosphere today, the Sun's ultraviolet radiation is largely absorbed by oxygen to form an ozone layer, which absorbs almost all short-wave (high-energy) ultraviolet rays. This was not the case 4500 m yr ago: most of the Sun's energetic ultraviolet radiation penetrated to the surface of the Earth. Great amounts of photochemical energy were then available to interact with all components of the atmosphere and their products. Solar short-wave ultraviolet radiation was available from the beginning up until about 2000 m yr ago, when free oxygen was biologically produced. This radiation was of profound importance to the origin and evolution of life over the course of the first 2500 m yr.

Other sources of energy, far greater in primeval times than at present, were electrical discharges and heat. Volcanic activity was the predominant surface feature of the ancient Earth. Heat from laval outflows into ponds and lakes created a variety of rapidly changing environments and provided sufficient energy to drive many synthetic reactions, as Chapter Five will discuss. Rainstorms were intense and frequent, and the sky was clouded with volcanic ash. Thunderstorms took place almost continuously; massive amounts of energy were released as electrical discharges, both in the atmosphere and from clouds to the surface.

Debris from planet formation was still abundant. Objects ranging in size from dust specks to massive aggregates were still being trapped by the Earth's gravitational field, causing continuous meteoroid and meteorite bombardment. The considerable energy from the heat released upon meteorite descent and impact and from atmospheric shock waves interacted with the terrestrial atmosphere and surface.

In a cosmic sense, the matter of the primitive Earth was in part quite young. Short-lived radioactive elements were decay-

TABLE 4-2

Major energy sources of the primitive Earth.

SOURCE	ENERGY (CAL/CM²/YR)	RELATIVE ENERGY
Solar short-wave ultraviolet radiation	570	712
Electrical discharges	4	5
Radioactivity	0.8	1
Volcanic heat	0.13	0.16
Meteorite impact	0.05	0.06

ing, raising the temperatures in the Earth's interior and contributing to volcanism. On the surface, "background" radiation was considerably higher than we might wish to experience. Radioactive decay energies in the form of beta particles, alpha particles, and gamma rays were a significant feature.

Table 4-2 gives a guess at the relative amounts of energy from these various sources. Clearly, the gases emitted from volcanoes, confronted with this gross abundance and variety of energy, must have undergone numerous and complex interactions and reactions. Chapter Five will discuss the outcome of this interplay between matter and energy.

As internal temperatures increased because of radioisotope decay and gravitational compaction, at about 4000 m yr ago the semimolten interior of the Earth underwent a drastic reorganization. Metallic iron and nickel, which earlier had been mixed with other minerals in the upper mantle, reformed to generate a molten iron-nickel core over which lay the new mantle and surface crust. Core formation seems to have taken place abruptly, as geological time is measured; it might have been the result of continued deep interior melting. Since this early forma-

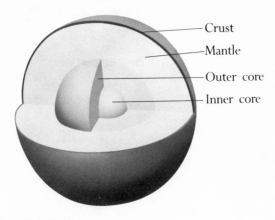

FIGURE 4-1
Core, mantle, and crust of the Earth.

tion of the core, the deep interior features of our planet have remained much the same for 4000 m yr (Figure 4-1).

Since core formation removed much of the metallic iron from the upper mantle, volcanic magmas thereafter presented a different composition, one in many ways similar to present-day volcanic magmas. The melt consisted mainly of metallic oxides and silicon oxides, such as MgO, SiO_2, FeO, and Fe_2O_3.

The pressure of oxygen has also been calculated for these melts; it is $10^{-7.1}$ atmospheres. Although this number is still very small, oxygen was some 100,000 times more abundant at this stage than before the Earth's core was formed.

This stage has been called the Hawaiian volcanic stage, since once core formation had occurred the upper mantle and crust would have contained the same molten minerals as they do now. Similarly, their associated volcanoes would have emitted the same gases then as now. This point of similarity provides a good test of theory, for we can calculate the expected volcanic gas ratios and compare them to actual measured volcanic gas compositions. Such comparisons show a comforting agreement.

What do these ratios tell us? In the Hawaiian volcanic stage, water is some 100 times more abundant than hydrogen;

TABLE 4-3

Later volcanic outgassing atmosphere of the primitive Earth.

MAJOR GASES	MINOR GASES	TRACE GASES
Water vapor (to oceans)	Sulfur	Methane
Carbon dioxide (to oceans)	Carbon monoxide	Hydrogen sulfide
Nitrogen		Hydrogen
Sulfur dioxide		

carbon dioxide is 40 times in excess over carbon monoxide; sulfur dioxide is the dominant form of sulfur; methane is virtually absent; and nitrogen continues to be emitted steadily. Table 4-3 gives the composition of volcanic gas at the moment of ejection.

During the course of the first 500 m yr of the Hawaiian stage, volcanic activity was still quite high, meteorite impacts were declining somewhat, the oceans were increasing in size and in salt content, and electrical discharges remained as frequent. Most of the energy forms of the preceding era still prevailed, including solar short-wave ultraviolet radiation. An important aspect of this phase is that the initial gas composition of the atmosphere, which reacts with the varied forms of energy, has changed. Of the major gases, water condenses as rain, much of the carbon dioxide dissolves in the rapidly building oceans, and nitrogen accumulates in the atmosphere. During this phase of rapid ocean building, the atmosphere probably consisted largely of nitrogen gas, with carbon dioxide at its present-day value of 0.03% and only traces of other gases. What little hydrogen was present would rapidly combine with sulfur dioxide to make hydrogen sulfide or would be lost by escape from the upper atmosphere.

This stage of atmospheric evolution began 4000 m yr ago and continued for some 1500 to 2000 m yr; it ended only when biological oxygen production began. Throughout this vast time span, the physical environment gradually changed: volcanic activity decreased, oceans increased in volume and salt content, meteorite bombardment declined. But the Sun continued to shine with all its energetic ultraviolet radiation, and biological life evolved—life that did not produce oxygen from photosynthesis. The atmosphere was much as it is now, with one outstanding exception—no free oxygen was present.

The oldest known sedimentary rocks are from this era, indicating the buildup of stable oceans. The first known sedimentary rocks are some 3800 m yr old. Chemical clues found within these give support to our picture of the environment at this time. For ages greater than 4000 m yr only crystalline rocks are available, indicating that shallow ponds and lakes were quasi-permanent, with insufficient time to collect, compact, and mix volcanic ash, crushed fragments of rocks, and other precipitates to form sedimentary rocks.

The third and last stage of our atmospheric evolution story, called biological, is characterized by the buildup of free oxygen. Oxygen is a very reactive element: it combines with various elements in rocks, and it reacts freely with volcanic gases such as carbon monoxide, sulfur dioxide, and hydrogen. Without a biology, only a faint trace of oxygen could persist in the atmosphere.

In the absence of biological life, trace levels of oxygen are constantly being produced and used. Production results from the photodissociation of water. High in the atmosphere, water vapor interacts with solar ultraviolet radiation:

$$2H_2O \xrightarrow{\text{solar ultraviolet radiation}} O_2 + 2H_2$$

At these heights the hydrogen gas escapes from the Earth, and the oxygen is free to react, usually by weathering the less oxidized gases and minerals. A constant production and use of oxygen thus take place. Some 2×10^{12} grams/year of oxygen are produced in this fashion. This number appears large until we consider that present-day photosynthesizing plants have produced some 181×10^{20} total grams per year of oxygen. In fact, although photosynthesis produces so much oxygen, only some 12×10^{20} grams remains in a steady state in our present atmosphere. The rest, some 171×10^{20} total grams per year, is used in weathering processes. The residue that cannot immediately be used in weathering constitutes 20% (by volume) oxygen level currently found in the atmosphere.

The free oxygen state of atmospheric evolution began some 2000 m yr ago, when photosynthesizing organisms evolved a way of using visible sunlight as an energy source and water as an electron donor. Oxygen is a byproduct of this process. The history of free oxygen presence above trace photodissociation amounts can be determined by examining the oxidation state of iron in sedimentary rocks. Rocks from 2000 m yr ago to the present are found to be relatively much more oxidized than are older rocks.

Accumulation of free oxygen in the atmosphere probably progressed in a linear fashion from 2000 m yr ago until 600 m yr ago, when complex multicellular life forms evolved. As photosynthetic free oxygen accumulated in the atmosphere, it would have absorbed solar ultraviolet radiation and formed the ozone (O_3) layer. As a consequence, solar ultraviolet radiation very suddenly ceased to penetrate to the surface of the Earth. Although this radiation drives chemical reactions, it can also degrade organic compounds and biological systems. The establishment of the ozone shield was one of the most significant environmental steps, since it allowed a terrestrial biology to develop.

CHAPTER FIVE

CHEMICAL EVOLUTION

In the cauldron boil and bake
Eye of newt and toe of frog,
Wool of bat and tongue of dog,
Adder's fork and blindworm's sting,
Lizard's leg and howlet's wing . . .

SHAKESPEARE, *Macbeth*

The Urey–Miller spark discharge experiment. In this experiment a 2-liter flask holds a mixture of methane, ammonia, and water vapor, which is exposed to a continuous electrical discharge between the two electrodes. Note the heavy deposition of polymeric material on the glass surface. (Photo courtesy of NASA.)

At the Earth's beginning some 4500 m yr in the past, shallow lakes and ponds of fresh rainwater, murky with volcanic ash, were forming over its crust. Surface temperatures were about 25°C—room temperature. Volcanic activity, thunderstorms, meteorite falls, and energetic solar ultraviolet radiation were all intense; energy of all forms abounded. The major gases in the atmosphere were nitrogen, carbon monoxide, hydrogen, and water vapor.

These gases could not have remained unchanged under these new conditions. Every chemical reaction is reversible and environment-dependent. For example, in the reaction

$$CO \quad + \quad 3(H_2) \longrightarrow CH_4 \quad + \quad H_2O$$

| carbon monoxide | hydrogen (in excess) | methane | water |

the amount of each reactant present and the temperature of the environment determine how much product is formed. One can calculate that at room temperature, if hydrogen is abundant, practically all the carbon monoxide will quickly be converted to methane.

Similarly, in the nitrogen reaction

$$N_2 \quad + \quad 6(H_2) \longrightarrow 2(NH_3)$$

nitrogen   hydrogen   ammonia
(in excess)

almost all the nitrogen would be converted to ammonia.

Analysis of these reaction equations, for which exact reaction rates are known, leads to a prediction that the resultant atmosphere would consist of methane and ammonia. This is far too simplistic an approach, however, and early rocks give no evidence that an ammonia-methane atmosphere ever existed.

If methane were abundant, this gas would quickly have been further converted into all kinds of hydrocarbons and oils, and these products would be manifest as large deposits of chemically modified hydrocarbons and graphite within early rocks. But there are no such deposits.

If ammonia were present, it would quickly have dissolved into all bodies of water to make ammonium hydroxide, a strong base. The ponds and lakes would have been far more alkaline than they are now. Rocks from this early time contain silicates that could not have precipitated under such alkaline conditions. Indeed, if any ammonia were in the atmosphere, it would have been decomposed rapidly by solar ultraviolet radiation.

Are our equations wrong, then? No, they just do not address the facts. The difficulty is that each chemical reaction equation represents a single clean, isolated set of observations. The reality is entirely different because all the atmospheric components were in fact mixed together, interreacting with each other and with products in all possible combinations and permutations. Figure 5-1 outlines this more complex aspect of nature.

The essential observation is that in the real world, no one specific isolated reaction can be set aside peacefully to go to

Gases upon ejection from primeval volcanoes

$$a(CO_2) + b(CO) + c(N_2) + d(H_20) + e(H_2) + f(S)$$
$$+ g(H_2S) + h(CH_4) + i(SO_2) + \cdots$$

$$\rightleftharpoons$$

react in the presence of various abundant
forms of energy to form a variety of
small reactive organic compounds:

$$j(HCN) + k(HCO_2H) + l(CH_3CO_2H) + m(HN(CN)_2)$$

hydrogen cyanide — formic acid — acetic acid — dicyanamide

$$+ n(H_2NCH_2CO_2H) + \cdots$$

glycine

$$\rightleftharpoons$$

which react with various abundant
forms of energy to yield more complex organic compounds:

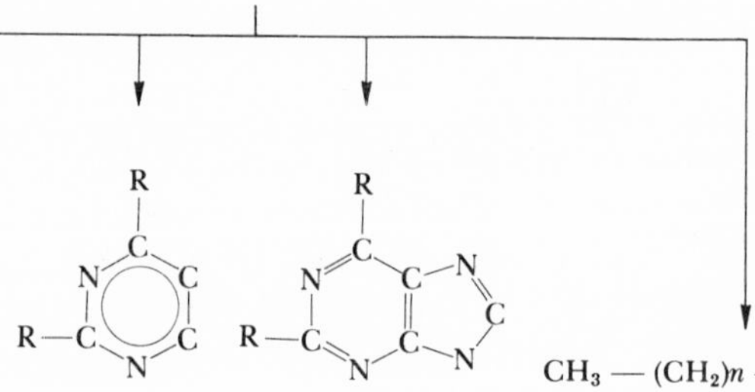

$$R - \underset{\displaystyle NH_2}{\underset{|}{C}} - CO_2H \qquad CH_3 - (CH_2)n - CO_2H$$

amino acids (R = various organic groups) — pyrimidines (R = various groups) — purines (R = various groups) — fatty acids

FIGURE 5-1
Gases upon ejection from primeval volcanoes. Small letters represent the amounts of the various components.

completion. The number of possible interactions accumulates rapidly, leading to formation of a great variety of quite complex products in short periods of time (measured in years). Some of these complex products are the "tarry intermediates" that Harold Urey suspected might form from such starting materials. The rest of the products constitute a wide range of organic molecules basic to the dawn of the origin of life.

How can we conduct meaningful simulation experiments if we are unsure of the composition of the starting atmosphere? The exact composition of the "real" primitive atmosphere was continuously changing; it was a kaleidoscopic mixture of fresh volcanic gases and volatile reaction products. At this time a single gas—hydrogen—directed the general features of all products formed, and we know that hydrogen was a major product of volcanic outgassing. And as we shall see, when any gas mixture containing hydrogen interacts with the many possible energy sources, essentially the same products are formed.

The fates of primeval volcanic gases depended on two factors. The first is chemical evolution—the reaction and interaction rates of all gases and their various products with all forms of energy. The second is the escape rate of hydrogen from the upper atmosphere. Earth's gravity is insufficient to retain hydrogen from the outer reaches of the atmosphere at prevailing temperatures. Thus the atmosphere will lose hydrogen at a constant rate and will evolve into the products of chemical evolution.

Since 1897 many simulation experiments have been performed. All of them point to the conclusion that energy interacts with primeval gases to form a large set of randomly synthesized organic compounds. Furthermore, the set of small organic compounds necessary for biological evolution is found within this larger, random nonbiological set. We will describe several classic experiments in chemical evolution that support this general conclusion.

In 1953, Stanley Miller and Harold Urey, at the University of Chicago, devised a straightforward experimental system in which methane, ammonia, and water were subjected to an electrical discharge. Although spark discharges are not as representative of the bulk of primeval energy as solar ultraviolet radiation, the latter is exceedingly difficult to generate and work with. However, chemical bonds can be made and broken by either form of energy.

The Miller–Urey apparatus is depicted in Figure 5-2. Gases such as ammonia and methane are bled into the reaction chamber after it is evacuated of air. Boiling water provides water vapor, and the condenser serves to circulate the gas mixture through the vessel where the spark discharge electrodes are located. After continuous sparking at 60,000 volts for several days—an energy input equivalent to that of some 50 m yr on the primeval Earth—the water phase was subjected to chemical analysis for synthesized organic compounds.

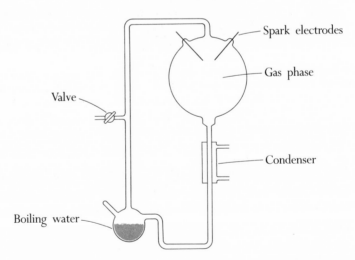

FIGURE 5-2
The Urey–Miller spark discharge flask.

TABLE 5-1

Types and yields of simple organic compounds obtained from sparking a mixture of CH_4, NH_3, and H_2. (Yields are relative to formic acid.)

COMPOUND	RELATIVE YIELD
Glycine	270
Sarcosine	21
Alanine	145
N-methylalanine	4
Beta-alanine	64
Alpha-amino-n-butyric acid	21
Alpha-aminoisobutyric acid	0.4
Aspartic acid	2
Glutamic acid	2
Iminodiacetic acid	66
Iminoacetic-propionic acid	6
Lactic acid	133
Formic acid	1000
Acetic acid	64
Propionic acid	56
Alpha-hydroxybutyric acid	21
Succinic acid	17
Urea	8
N-methyl urea	6

The separation, identification, and quantification of a mixture of many different kinds of organic compounds is a long and laborious task. But it recovered an extensive variety of small organic molecules. Among these were amino acids, which are found in biological proteins, and other small compounds such as lactic acid, which is involved in biological sugar utilization. Table 5-1 lists some of these compounds. An immediate observation is that many biologically important small organic compounds were formed with great ease and in abundance. Also found are related small organic compounds that are not universally used in our

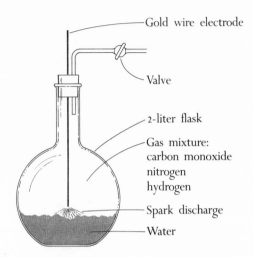

Gold wire electrode

Valve

2-liter flask

Gas mixture:
carbon monoxide
nitrogen
hydrogen

Spark discharge

Water

FIGURE 5-3

*A modified Urey—Miller spark discharge
experiment. A 2-liter flask containing a
small amount of liquid water is connected to
a single electrode and to a vacuum valve at
its top. Through the valve, air is evacuated
and a gas mixture introduced. A
high-voltage coil (30,000 to 60,000 volts)
connected to the single top electrode
discharges onto the water surface. After
several days of sparking, the reaction vessel
can be connected directly to instruments to
measure the kinds of volatile products
formed, or it can be opened and the
nonvolatile, water-soluble products can be
subjected to chemical analysis. Two days of
sparking represent an energy input into the
system comparable to some 40 million years
on the surface of the primitive Earth.*

biochemistry, as well as a large number of highly reactive, very simple compounds, such as hydrogen cyanide, formic acid, acetic acid, urea, and cyanamide.

The general course of reactions such as these is:

Primeval gases $\rightleftharpoons$ Reactive intermediates $\begin{array}{c} \nearrow \text{Polymeric material} \\ \\ \searrow \text{Organic compounds} \\ \text{(biological subset)} \end{array}$

driven by the energy of the spark discharge.

Since 1953, numerous workers have performed similar experiments, varying input gases and energy sources (Figure 5-3); they have refined the many methods of organic chemical analysis. As long as the starting mixture of gases contains carbon (in some gaseous form), nitrogen, water, and some hydrogen, the general course of the reactions is consistent. Reactive intermediates, a wide set of small organic compounds, and polymeric material are the results. The process is not at all energetically efficient, but there is no reason why it has to be, since so much energy of varied types was then available.

Spark discharges in gas mixtures represent only one of the forms of energy input in the primeval environment. Other forms are ultraviolet radiation, heat (Figure 5-4), shock waves, beta and gamma particle irradiation (Figure 5-5), and cosmic rays. Experiments have been performed using these forms of energy and various primeval gas mixtures.

Edward Ander's group at the University of Chicago has studied Fischer–Tropsch type reactions, which involve heating gas mixtures to high temperatures (600 to 900°C) in closed containers in the presence of various metallic ore catalysts. These experiments simulate reactions between the primitive

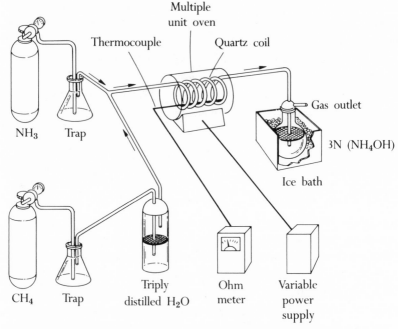

FIGURE 5-4

A thermal synthesis experiment. Gases such as methane, ammonia, and water vapor pass through a tube contained within a furnace. The tube contains powdered metallic catalysts and is heated to temperatures of 900°C. The gases react, yielding nonvolatile products that are trapped at the exit of the tube.

atmosphere and frequent lava flows. A great and complex variety of organic molecules has been isolated from such model reactions. Not only are amino acids found, but hydrocarbons, sugar precursors, nucleic acid bases, fatty acids, and reactive intermediates are also present. Table 5-2 lists results compiled from a number of such experiments.

A. Bar-Nun has demonstrated that shock waves comparable to those triggered by meteorite falls can interact with gase-

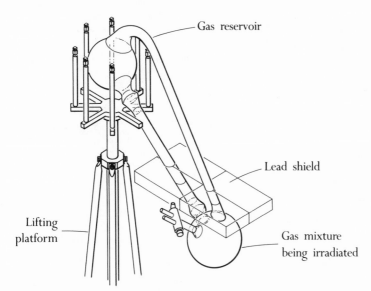

Gas reservoir

Lead shield

Lifting platform

Gas mixture being irradiated

FIGURE 5-5

Apparatus for gamma radiation of methane, ammonia, and water. Methane, ammonia, and water vapor are admitted to the apparatus through the valve in the lower flask, which is placed in the path of a gamma source and shielded with lead bricks. The upper flask acts as a source of further gases and as a radiation-free site in which products can accumulate without being destroyed by radiation.

ous components to effect the formation of amino acids and reactive intermediates.

Simulated solar energetic ultraviolet radiation, generated by a giant argon plasma arc beamed through a magnesium fluoride window onto mixtures of primitive gases, results in the synthesis of amino acids, polymers, and many other compounds.

Radioisotope decay energies can be simulated by use of an electron accelerator. Melvin Calvin's group at Berkeley irradiated mixtures of methane, ammonia, and water and found reactive

TABLE 5-2

Some organic compounds formed from Fischer—Tropsch type reactions, in which CO, H_2, and NH_3 gases are heated to 900°C with metallic catalysts.

"BIOLOGICAL" AMINO ACIDS	"BIOLOGICAL" NITROGEN BASES	FATTY ACIDS
Glycine	Adenine	All varieties from
Alanine	Guanine	C_{12} to C_{20}
Valine	Xanthine	
Leucine	Thymine	
Isoleucine	Uracil	
Aspartate		**HYDROCARBONS**
Glutamate		A wide variety
Tyrosine	"NONBIOLOGICAL"	resembling that
Proline	NITROGEN BASES	isolated from
Ornithine	Melamine	carbonaceous
Lysine	Ammeline	meteorites
Histidine	Cyanuric acid	
Arginine	Guanyl urea	

"NONBIOLOGICAL" AMINO ACIDS	SMALL REACTIVE MOLECULES
N-methyl glycine	Hydrogen cyanide
Beta-alanine	Formaldehyde
Alpha-aminoisobutyrate	Formic acid
Alpha-amino-n-butyrate	Acetic acid
Beta-aminoisobutyrate	Urea and related compounds
Gamma-aminobutyrate	Guanidine and related compounds

intermediates and a number of biologically related compounds, including adenine.

All these series of experiments point in the same direction and support the theory of chemical evolution. All show that interplay of all forms of energy with gaseous carbon, nitrogen, water, and hydrogen leads first to the synthesis of reactive intermediates. These intermediates then interact to form a larger

TABLE 5-3

An overall view of the products formed in chemical evolution experiments.

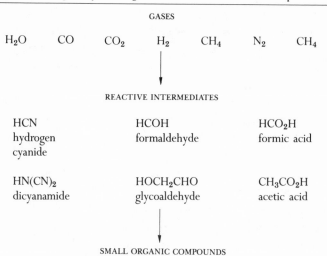

GASES

H_2O	CO	CO_2	H_2	CH_4	N_2	CH_4

REACTIVE INTERMEDIATES

HCN hydrogen cyanide	HCOH formaldehyde	HCO_2H formic acid
$HN(CN)_2$ dicyanamide	$HOCH_2CHO$ glycoaldehyde	CH_3CO_2H acetic acid

SMALL ORGANIC COMPOUNDS

Amino acids: several hundred, including the biological subset of 20

Fatty acids: all varieties from C_{12} to C_{20}

Hydrocarbons: a complex "random" mixture

Nitrogenous bases: uracil, cytosine, thymine, adenine, guanine, xanthine, hypoxanthine, triazines

Other small organic compounds: urea and derivatives, guanidine and derivatives, acids and diacids

Polymer

---

set of biological and biologically related organic components. No single experiment is representative of the primitive Earth, but all of them together might be. Note also that polymer seems always to be found along with smaller organic molecules. Table 5-3 lists a sampling of reactive intermediates and small organic compounds formed from these kinds of reconstruction experiments.

In studying the abiologic (model) synthesis of amino acids in great depth and detail, Stanley Miller's group at La Jolla and James Lawless's group at NASA Ames Research Center have found that *all* possible kinds of amino acids are formed in abiologic syntheses. This set contains equal amounts of the mirror-image L and D forms; within this set is found most of the entire biological subset.

Thus chemical evolution—the idea that all common small organic molecules used by our biology will be formed from primeval simpler components interacting with various energy sources—derives support from these experiments.

Where do all the newly synthesized organic materials go once they are formed on the primitive Earth? Some volatile ones, such as HCN and formaldehyde, remain largely in the atmosphere and to some extent in solution in ponds. Others, including amino acids and nucleic acid bases, are mainly dissolved in the water of ponds. Hydrocarbons and long chained fatty acids float as a scum on the water's surface. Since solar ultraviolet radiation and all other energy sources continue their input, not only will synthesis of organic compounds continue, but their degradation will also occur. This leads to a steady state—a cycling of components from gases to organic compounds back to gases:

$$H_2 + CO + N_2 + H_2O \underset{\text{destruction rate}}{\overset{\text{production rate}}{\rightleftharpoons}} \text{Organic compounds}$$

Both rates are mediated by energy input.

Calculations to predict the amount of, say, amino acids that would be dissolved in the pond at steady state give a very low abundance, about 10 millionths of a gram per liter for any one amino acid. Such a pond certainly cannot represent an "organic soup" rich in all small precursor molecules, on the brink of bursting forth with newly evolved life. It is rather an extraordinarily dilute brew of organic compounds with some dissolved metal ions. Some of the ever-present polymeric material sinks to the shallow bottom, and the surface is overlaid with a slight oil-like scum.

This might seem a grim stage setting for the origin of life, for most biological processes require a more concentrated milieu than our pond. However, numerous processes can occur to permit the evolution of life from ecological niches such as these; Chapter 7 will explore these processes.

CHAPTER SIX

A DEFINITION
OF LIFE

'Twas brillig, and the slithy toves
Did gyre and gimble in the wabe;
All mimsy were the borogoves
And the mome raths outgrabe.

LEWIS CARROLL

View of the Earth from an orbit about the Moon. (Photo courtesy of NASA.)

*L*ife is an elusive thing to define fully and completely: it wiggles away whenever approached. One can wax phenomenological and prepare a list: a "living" creature moves, excretes, eats and metabolizes, reproduces, grows, and so forth. The problem with this approach is that there are always disquieting exceptions to the list that threaten the basic definition. The definition itself, an arbitrary list of attributes, is hardly fundamental. Problems arise when one encounters a virus, a growing crystal, a bacterial spore, or a complex, artificially intelligent computer.

Lars Onsager and Harold Morowitz have developed a unique and successful approach to this problem; we will develop their approach here. One of the most basic problems involved in defining life is that we are so close to it and so much a part of it that its essential features are obscured in its overwhelming details. In fact, if one mentally steps off our green-blue planet only to look behind, the most essential single thread tying all living systems together is our balanced ecology. "Life" is an ecological property; it is only an individual property for a brief flash of time.

An ecology is a oneness of varied complex biological systems, all making continuous and subtle trades with one another, joined by the common goal of using to the utmost all that the present environment has to offer in terms of energy and raw materials. An ecological approach to define life removes us from the obscurities of detail to reach for fundamentals.

Let us phrase and then examine an Onsager–Morowitz type of definition of life, legalistic though it may seem:

Life is that property of matter that results in the coupled cycling of bioelements in aqueous solution, ultimately driven by radiant energy to attain maximum complexity.

Certainly this definition will not make us unique, but are we really? This approach has several important advantages. Every facet of it can be approached by experimental test. It is based upon a newly forming general theory of biology that states that an energy flow through a system acts to organize that system, to make it more complex. It is an ecological, whole-oriented definition: a cycling of bioelements cannot exist without primary producers, users, and scavengers. An ecology of the whole is a sum of all the various doings of plants, animals, and microorganisms.

If all plants were by magic removed from our planet, what would happen? Very rapidly all the free oxygen in the atmosphere would disappear, the ozone shield would break down, and most, if not all, life would cease to exist. If all the microorganisms of the world were to disappear, the recycling of biological elements would halt; nothing could rot to provide future sustenance for succeeding generations. We all depend upon all.

Accepting the definition of life just presented, we can now rephrase the question "How did life come to exist?" more specifically:

How did aqueous coupled bioelemental cycles driven by radiant energy to become more complex come to exist?

"Bioelemental" is the first facet of the question that requires examination in greater detail. The bioelements are carbon, hydrogen, nitrogen, oxygen, phosphorus, and sulfur

(CHNOPS). Why these and not others? Why not a silicon-based life breathing fluorine? Each of the bioelements is cosmically abundant and uniquely suited for its role. With the possible exception of phosphorus, these were the gases that were originally the first line of interaction of matter with radiant energy. What was present—gases of C, H, N, O, and S—had to react when and how it could, and thus we are as we are. There would be no reason to expect anything different in any other terrestrial-size planet within a proper temperature region about its star. Why not silicon? Because this element is occupied full time in crystalline rock building. Silicon has no dominant gaseous forms to interact with radiant energy. Whatever organism we analyze for its elements, our results are essentially the same: more than 90% of its mass is CHNOPS.

"Aqueous"—water-based—is another necessarily ubiquitous biological property. Dormant biological systems, such as bacterial or algal spores or brine shrimp eggs, are not functioning and contain little water. All functioning systems consist mostly of water. All biochemical reactions occur within water because most organic compounds of the biological subset are water soluble. Those others that are volatile become dispersed in the atmosphere to react and be degraded by radiant energy. Those that are water insoluble sink to the pond's bottom, removed from further interactions. What remains to react in aqueous solution are dominantly water soluble organic compounds. This is the heritage of our primeval ecology.

"Coupled cycles" requires some explanation. Nonbiological cycles of matter, all driven by radiant energy, exist in every form on Earth and elsewhere. For example, oxygen can be formed by the photodissociation of water; it can then react with volcanic hydrogen to appear again as water. Biological systems manifest material cycles too; but the difference is that all bioelemental cycles are coupled to one another. We can see this in an organism's growth—a regular, orderly increment in size and com-

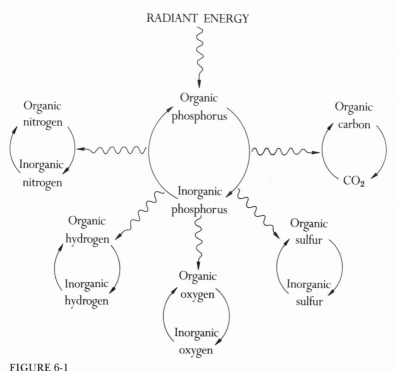

FIGURE 6-1

Coupled Onsager bioelemental cycles. Radiant energy drives the phosphorus cycle. All the other bioelemental cycles are keyed to this cycle and are driven by it in specific turn ratios.

position of all components. The same bioelemental coupling takes place in the ecology on a planetwide scale (Figure 6-1). The phosphorus cycle is the cycle driven by radiant energy. For every turn of this cycle, all other cycles are driven in their own specific ratios.

"Driven by radiant energy" means that visible-frequency solar energy ultimately drives all biological systems. Neither the biological phosphorus cycle nor any other will turn spontaneously; they all must be driven by solar energy—directly as in photosynthesis or indirectly as in consumption of plants by an herbivore. Energy, like time, is a complex concept that we can measure only by its interplay with matter: it vanishes at the moment we approach the essence of its nature. The fact remains that energy brings about transformations of states of matter. When these transformations are coupled, as they are for the bioelements, we have a biology, as the totality of our experience and experiments have told us.

Probably the most subtle, vital, intuitively obvious, and confusing aspect of this definition of life is encoded in the phrase "to attain maximum complexity." "Complexity" was chosen over "information" because the latter term has too many varied meanings to different beholders. As Darwin has shown, natural selection will choose from any large population only that subset most capable of maximizing its numbers in its ecological niche. The operation of such selection is based on genetic variation inherent within the population and on the fact that any population potentially reproduces itself to gross excess. But more is involved in the issue of complexity, for with these tenets alone one could imagine a world of superevolved microbes capable effectively and rapidly of maximizing every potentially available niche to the exclusion of all else. Yet humans, plants, and bugs do exist and are of marvelous complexity. Thus, niches must abound which are inaccessible to microbes, representing a haven for complexities of form—for multicellular organisms. A general biological theory should in time be capable of delineating fully these niche attributes and of coping adequately with the complexity problem. Present theory cannot; in fact it is now problematic whether one can usefully measure the "information" content of an organism's DNA. But one niche attribute stands out: the conjoint properties of size and time on the biological scale.

Microbial cells are small, limited basically by the physics of diffusion of requisite organic compounds from without: volume increases in proportion to the cube of linear size, while the surface area available for diffusion only increases in proportion to the square. In addition, diffusion processes require time. More complex, larger eucaryotic cells developed from aggregates of microbial cells. The critical innovation involved in aggregates of primitive bacterialike cells was an increase in internal surface area by use of many systems of membranes. A necessary corollary was an increase in the time intervals required for cell division. This newer, more complex cell could process more raw material faster and occupied a different time niche: more molecular events could take place inside a more complex cell within a shorter time. The opening of new niches required more than the evolution of greater complexity in developing biological systems. Also necessary were environmental fluctuations that could interact with a biological system's tendency to reproduce to excess and thus select the more fit.

Free oxygen became available largely through the metabolism of photosynthetic blue-green algae. In response to the presence of free oxygen, complex large eucaryotic cells evolved, since these cells could use oxygen to increase local biological energy production rates by a large factor. This was probably the most important single development in the evolution of multicellular life forms.

Although biological theory cannot now adequately predict from first principles why the complexity of energy-driven systems must increase, simple experiments of physics do indicate that this is so.

A classic thought experiment is to expose one side of a closed container to a point source of heat. The heat source is analogous to the flow of radiant energy. Gas molecule concentration within the box will become lower near the heat source and higher away from it. The overall order of the system has thus increased, since gas molecule concentration within the box is displaced from its more random state as long as energy flows.

Because of the inadequacy of current biological theory, a

very real difficulty remains when one actually attempts to measure order of this type in biological systems. For now, let us intuitively accept the hypothesis that order (or information, or complexity, or nonrandomness) is a definite characteristic of any biological system, and that when a large flux of high-grade radiant energy flows through either a biological or a nonbiological system, a small part of it reappears as order, while the rest reappears as low-grade heat.

Coupled bioelemental cycles imply that materials are used in common by all biological systems, and this is found to be so. All living things have similar biochemistries, which are based upon a surprisingly small number of organic compounds. Of the hundreds of thousands of organic compounds that could be made, only some 120 are used ubiquitously in our biochemistry. These include 20 amino acids, 4 lipids, 5 nucleic acid bases, 20 vitamins, and a miscellaneous group of small molecules such as organic acids, phosphate, and water. The small size of the set of pivotal organic compounds common to all living systems permits the various trophic levels of ecology to operate efficiently. Consider the impossibility of eating a tiger (or the converse) if our basic small molecule subsets were different.

From this small set of compounds the bulk of the nonwater mass of all organisms is formed. The number of classes of polymers is smaller still. Proteins, polysaccharides, lipids, and nucleic acids constitute the bulk of the nonwater mass of every functioning biological system. Again, the simplicity and universality of the polymeric structures of biological systems reflect the necessity of a common trophic level of ecology.

If an engineer were to design and build an extraordinarily complex computer, it would no doubt consist of arrays of basic units. Similarly, all biological systems possess a fundamental unit, the cell, which is in all cases bounded by a membrane. The size range of biological cells is quite narrow. Bacteria, or procaryotes, have smaller cells ranging from ½ to 2 μm across (1 μm $= 1 \times$

10^{-6} m). Eucaryotes—cells with nuclei and organelles—are more complex and larger, ranging in size from 5 to 20 μm. A cell isolates a portion of the aqueous environment within a phase boundary, the membrane, with the object of modifying that local environment of biosynthesis.

The physics of diffusion of small molecules in water sharply limits the maximum size cells can attain. The minimum size that cells can have is fixed by the number of functions a cell must perform to reproduce itself. A virus is not a functioning biological system; by our definition it is not living, but represents merely a protein-coated hypodermic syringe loaded with nucleic acid. It is a product of cells.

All cells have the same kinds of genetic and protein-synthesizing apparatus, which will be discussed in Chapter Nine. One of the most difficult facets of the origin-of-life question is to explain how such a genetic and protein-synthesizing apparatus arose. We will make an attempt to do so in Chapter Ten.

In all biological systems, proteins are the molecular machines that permit exact biosynthesis of polymers and enable energy and material transformations to occur quickly. Very few, if any, proteins function alone; rather, they are complexed with metallic or other ions. Specific instances are the occurrence of magnesium in chlorophyll and of iron in hemoglobin. These metallic elements are vital to functioning biological systems, although they are quite different from the bioelements. For example, the metals are ionically rather than covalently bonded. Although not a facet of biological structure, their presence is crucial to biological function.

Even such a brief review of the most general features of biology shows clearly that physical and chemical laws and processes set the boundary conditions for the size and functional nature of the cell. In addition, the restriction of a common trophic level imposed by an ecological system results in a universal biochemistry shared by all biological systems.

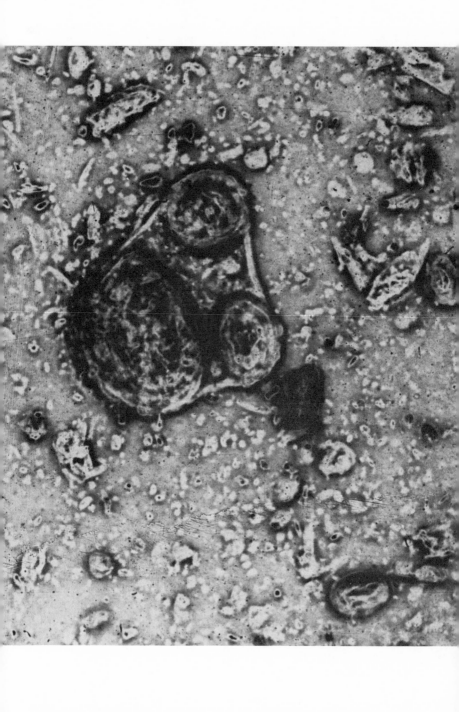

CHAPTER SEVEN

PROTOBIONTS

"There's no use trying," Alice said;
"one can't believe impossible things."
"I dare say you haven't much practice,"
said the Queen.
"When I was your age, I always did it for half
an hour a day. Why, sometimes I believed as
many as six impossible things before breakfast."

LEWIS CARROLL

Organic microstructures formed in quenched spark discharge experiments. This photomicrograph shows a droplet of water taken directly from a spark discharge experiment. Magnification is about 10,000 ×. About 60% of the input carbon in experiments such as these is recovered in microstructures. The larger cell-like objects are some 15 μm in diameter. Microscopic experiments on the larger structures indicate that they are phase-bounded; when the osmotic strength of the medium in which they are suspended is altered, they swell or shrink accordingly. Structures such as these are found in almost all kinds of spark discharge experiments in which a water phase is present, if the hydrogen content of the input gases is not excessive. (Photo courtesy of the Laboratory for Exobiology, University of Hawaii.)

"*P*rotobiont" is a name coined to represent an intermediate stage in the origin of life, between chemical evolution on the one hand and clearly biological forms that possess a genetics and are subject to Darwinian natural selection on the other. Some biological attributes were gained during this stage, and others were evolving.

A variety of hypothetical laboratory and conceptual model systems have been devised to explain the origins of cellular life. All seem plausible to some degree, and no doubt all of them and more were tried in the vast 1500 m yr of time between 3000 and 4500 m yr ago.

A. I. Oparin and co-workers propose a coacervate droplet composed of mixtures of colloidal particles as a model protocell. Colloidal particles can be long, largely unspecific macromolecules which have water bound to them and which will coalesce into cell-size coacervate structures under conditions of appropriate pH, salinity, and temperature. Coacervate mixtures can easily be made in the laboratory: for instance, dilute solutions of gelatin (a protein) and gum arabic (a polysaccharide) can be mixed. The solution remains clear until the pH is made acid; then coacervate particles settle out. In this case a low pH value is necessary for gelatin and gum arabic to interact to form cell-size structures. Coacervates can also be made from polyadenylic acid (a polynucleotide) and histone (a protein). Many other polymeric molecules will also interact in this fashion, although each kind of polymer has a specific set of environmental conditions under which it will form coacervates with other polymers.

These model systems are of great value in that they permit us to study specific synthetic and biochemical reactions in cell-size systems that partly sequester the local environment, allowing it to become different from the exterior. For example, Oparin has demonstrated that coacervates made of ribonucleic acid and histone, when given the enzyme polynucleotide phosphorylase in the presence of adenosine diphosphate, will synthesize the polynucleotide polyadenylic acid. This alone would happen without coacervates, but the essential point is that if sufficient histone is about, the coacervates will grow in volume and perhaps split mechanically in two. Other coacervates, provided with chlorophyll, will transport electrons from a donor dye to a receptor dye, modeling an aspect of photosynthesis.

Model coacervate systems use biologically made polymers: proteins, nucleic acids, and polysaccharides. Extrapolation to the world of the lifeless "warm little pond" requires accumulations of randomly made polymers which could coalesce into similar structures. Given many coacervates, each composed of random polymers, those that might slowly synthesize more similar polymers could grow and become selected for over a long time. Ultimately, in succeeding rounds of selection, they could develop a metabolism and a genetics.

I think this model is wrong, although it has been of immense use. To make coacervates in the laboratory requires quite high concentrations of polymers. But primeval ponds contained a decidedly dilute soup of small organic compounds. Hence the dilute small precursors must cross the first concentration gap to react and form polymers. The resultant dilute polymers must cross a second concentration gap to form coacervates. Finally, the coacervates themselves must cope with a most dilute solution of organic compounds to effect further coacervate synthesis. We will face this problem of the concentration gap again and again.

Hypothetically, there are ways to circumvent the concentration gap, but they all appear to be more wishful thinking than plausible facets of reality. For instance, since our ponds were numerous and shallow and interspersed among many grand volcanoes, they might evaporate. As they did so, water and volatile organic compounds, such as formaldehyde and hydrogen cyanide, were removed, leaving a soup most concentrated indeed. In some cases damp clays and ooze would have been all that remained until the next rain. No doubt this could and must have happened, but many difficulties still remain.

Salts such as sodium chloride are also present in ponds. Although primeval ponds contained far fewer salts than our present ocean, salt molecules still far outnumbered organic molecules. As evaporation proceeded, both would be concentrated. Consider some rough calculations. The salt concentration of a pond might be some 1.5 grams per liter (10% of the concentration of today's oceans), and amino acid concentration, of all some 20-odd, might be about 200 millionths of a gram per liter. The ratio of salt molecules to amino acid molecules is thus some 10,000 to 1. To focus on the synthesis of a random protein, we imagine a chance collision of one amino acid with another. The problem is that 10,000 times more frequently, the amino acid collides with a salt molecule! The above example also ignores the fact that amino acids are but one subset of an entire nonvolatile suite of organic compounds present in the evaporating pool. Thus, our amino acid would be colliding with and reacting with other dissimilar organic compounds as well as with salt molecules far more frequently than with other amino acids. Even worse, as the pool evaporated, solar ultraviolet radiation would become a factor, degrading quite rapidly whatever was concentrated. Under special circumstances concentration does work in the laboratory, but creating coacervates in the real world is quite another matter.

Another model of protocells forwarded by Sidney Fox seems to suffer from the same practical problems of the concentration gap. Like Oparin's, Fox's model has been quite thoroughly studied, and it is worth surveying. Fox observed that if a mixture of dry amino acids was heated and then plunged into water, polymers called proteinoids could be isolated from the mixture. Proteinoids are randomly synthesized proteinlike molecules formed by driving away one molecule of water for every bond created between two adjacent amino acids. When proteinoid solutions are cooled, they become turbid. Through the microscope, cell-like spheres about 10 μm across are visible; each of these possesses a thick shell-like proteinoid membrane. Sometimes apparent buds and doublings are evident, as Figure 7-1 shows.

Easily formed in the laboratory, proteinoid is a random mixture, mostly of polypeptides, with polymers of quite large sizes. Enzymes that degrade biological proteins will partially degrade these proteinoids.

Fox and his group have shown that these random proteinoids possess a wide variety of low-level catalytic abilities. They reason that amino acids are concentrated in evaporating ponds and then are polymerized by the heat of lava outflows or by sunlight drying. After rainfalls, the proteinoids self-assemble into microspheres. These microspheres represent a population of protocells that are subject to selection for the catalytic abilities requisite for a primitive metabolism.

Proteinoid microspheres are easy to prepare—it's done in many high school laboratories. All that is necessary is to heat a chunk of lava with a gas burner, throw a spoonful of dry L or D amino acids on the hot lava, and wash resultant proteinoids off the lava with a cup of water.

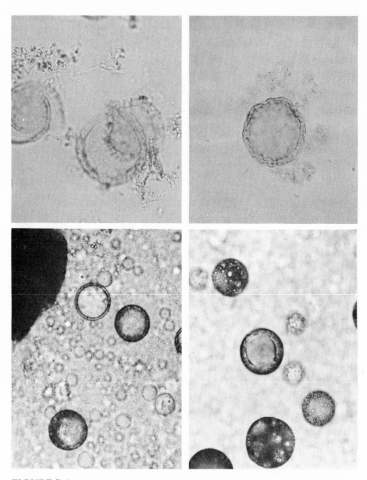

FIGURE 7-1

Proteinoid microspheres synthesized by Dr. Richard Young. Dry amino acid mixtures are condensed by heat or condensing agents to form proteinoids. Cooling of the proteinoid—water suspension results in the formation of microspheres.

The central question is where did all those pure, dry, concentrated and optically active amino acids come from in the real, abiological world? A further problem arises when we consider the nature of proteinoid microsphere boundaries. Cells possess a lipo-protein membrane, which is gossamer-thin and slowly permeable to many small molecules by diffusion. Proteinoid microspheres have a boundary made of grossly thick layers upon layers of partly hydrophobic proteins. This layer is so thick that it resembles a near-impermeable cell wall or spore coat more closely than a cell membrane.

In a somewhat different approach, J. D. Bernal proposed that the key to formation of polymers from small organic molecules is their adsorption, activation, and polymerization upon clays. Many different kinds of clay do indeed adsorb amino acids, nucleic acid bases, and sugars quite efficiently and selectively. In this model, clays circumvent the concentration gap somewhat, leading to high local concentrations of random biological-like polymers. At a later stage these polymers are assumed to self-assemble into protocells; and again, a primitive protocell population serves as a base from which those with synthetic and metabolic activity can be selected. One problem with clays is that materials upon them tend to be subject to ultraviolet radiation degradation. In addition, many polymeric materials are adsorbed more strongly than are small molecules and would not be available for further reactions. However, clays remain a useful concentrating mechanism, and they deserve further study.

If all the above protocell models face the major problem of the concentration gap, are there any alternatives? One option, which I personally favor, remains.

Whenever a mixture of primeval gases reacts with energy, more molecules are formed than just small organic molecules. *Invariably,* "polymeric" material is also recovered; it is usually the dominant form of the input carbon. This material is mainly hydrophobic and refractory to complete chemical characterizations. In a typical spark discharge reaction it is evident as a condensate on the walls of the reaction vessel and on the electrodes. Polymer is also formed when the spark is allowed to discharge through gas mixtures onto a water surface. In this instance organic material appears as a thin oily scum on the water's surface. When disturbed by motion, this scum separates from the surface to form spherules ranging in size from 1 to 20 μm in diameter, as well as more complex structures. The newly formed spherules possess a thin hydrophobic membrane enclosing other membranelike material and a portion of the aqueous environment. Figure 7-2 shows examples of the complexity of this polymeric material. The yield of this form of polymeric material is high.

Organic microstructures such as these apparently self-assemble from polymeric material on the water's surface; they then sink slowly to the pond's bottom. But tabulating the rate of formation of these structures, simply by counting them under a microscope at various times during a spark discharge reaction, leads to a most surprising observation: the number of microstructures increases exponentially with time. This indicates that one structure serves as a focus for the self-assembly of another. Biological populations of growing cells show the same autocatalytic growth kinetics. Other properties of organic microstructures will be discussed in later chapters. The essential clue their existence gives us is:

> *Protocells were formed from polymeric material within primeval ponds coincidentally with small organic compounds.*

Thus the formation of protocells does not require implausible concentrative mechanisms. Protocells were already present in abundance in the very first pond, created from the interaction of energy with primeval gases and scum. Then at the moment of

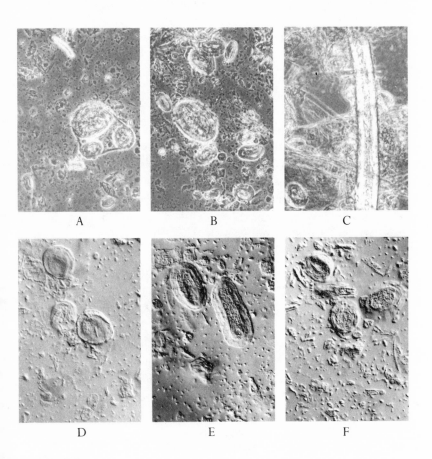

FIGURE 7-2

Organic microstructures formed by electrical discharge over a water surface. These photomicrographs of a droplet of the water after 20 hours of sparking were made at 900 × magnification using phase-contrast optics. The larger spheres are about 20 μm in diameter. Notice all the profusion of smaller bacteria-like objects and the abundance of thin membranous material comprising the larger structures. (Photos courtesy of the Laboratory for Exobiology, University of Hawaii.)

their creation they were shielded from severe ultraviolet radiation degradation by sinking to the pond's bottom.

Do ancient rocks give any direct evidence for the existence of protocells? How might such structures be distinguished from remains of truly biological cells?

Several methods of analysis have been developed in order to answer these questions. A sedimentary rock can be dissolved in a mixture of hydrofluoric and hydrochloric acids. The mineral matrix dissolves to leave unharmed fossilized organic material, which can then be visualized under the microscope. Rocks can also be cut into thin slices, polished, and microscopically observed for cellular microfossils. The work is tedious, and a great many samples must be scanned to record unambiguous microfossils. Collection and age dating of ancient rocks are formidable tasks in themselves. A further complication is that of deciding what criteria of biogenicity to apply to the observed fossilized organic material.

Bartholomew and Lois Nagy and their co-workers at the University of Arizona have described the oldest known organic microstructures, 3769 ± 70 m yr old, from the Isua micaceous metaquartzite from Greenland. Examples are shown in Figure 7-3. These microstructures are identical to those produced by spark discharges on water surfaces.

It is difficult to obtain older sedimentary rocks, since stable ocean systems did not form until about 4000 m yr ago. It does seem clear that these structures reported by the Nagys are protocell-like objects, at least in their morphology. The Isua structures also closely resemble organic microstructures recovered from South African Onverwacht sediments (3400 m yr old) and Fig Tree sediments (3000 m yr old).

The microfossil record thus indicates that protocell remains can be found in sediments 3000 m yr old and older. From 3000 m yr ago to the present, a change in microstructure complexity becomes apparent.

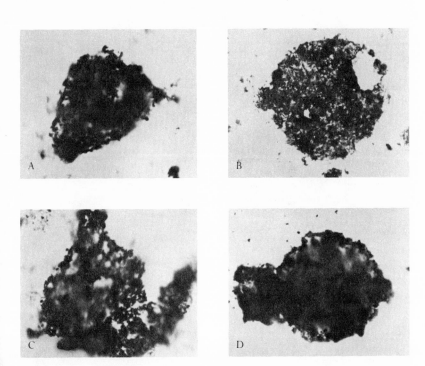

FIGURE 7-3

Organic microstructures seen in the Isua micaceous metaquartzite of Greenland by B. Nagy. All evidence points to an abiological synthesis of these structures. (Photo courtesy of B. Nagy.)

The oldest known undoubtedly biological microfossils were found in the Transvaal stromatolites (2200 m yr old) by Lois Nagy. These fossils are so similar to present-day filamentous blue-green algae that they compare to specific taxonomic groups (see Figure 7-4). Elso Barghoorn and William Schopf have reported rod-shaped bacterialike objects isolated from the Fig Tree series sedimentary rocks of South Africa and dated at 3000 m yr. These rod-shaped fossils could be protocell remains or bacteria. Criteria for biogenicity are sorely needed.

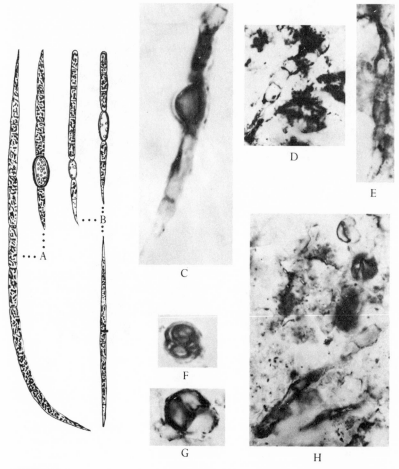

FIGURE 7-4

Unambiguous fossil remains of blue-green algae, found by Lois Nagy in the Transvaal stromatolite of age 2200 m yr. (Photo courtesy of L. A. Nagy.)

Stromatolites are obvious banded and sculpted deposits of carbonates in sedimentary rocks; they occur as the result of blue-green algal photosynthesis and algal mat deposition (Figure 7-5). These unique structures in sedimentary rocks are the best

FIGURE 7-5

Stromatolites. These characteristic wavy, laminated carbonate depositions result from the trapping and precipitating actions of the mucilaginous sheaths of photosynthetic blue-green algae and some bacteria. Although microfossil remains are rarely preserved in rocks of this type, the very presence of such biologically induced rock formations is a vital clue to early biological activities. This sample is from the earliest known stromatolite, the Bulawayan group of Southern Rhodesia, which is dated between 3100 and 2700 m yr old. (Photo courtesy of NASA.)

unambiguous evidence currently available for the existence of algal life forms. The oldest known stromatolites occur in the Bulawayan Group of Southern Rhodesia, which is dated between 3100 and 2700 m yr of age.

These data suggest that an enormous time was available for protocell experimentation, starting with the first protocell of

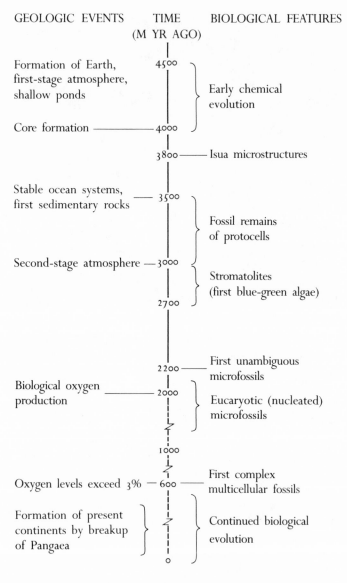

GEOLOGIC EVENTS	TIME (M YR AGO)	BIOLOGICAL FEATURES

Formation of Earth, first-stage atmosphere, shallow ponds — 4500

Core formation — 4000

Early chemical evolution

3800 — Isua microstructures

Stable ocean systems, first sedimentary rocks — 3500

Fossil remains of protocells

Second-stage atmosphere — 3000

2700

Stromatolites (first blue-green algae)

2200 — First unambiguous microfossils

Biological oxygen production — 2000

Eucaryotic (nucleated) microfossils

1000

Oxygen levels exceed 3% — 600 — First complex multicellular fossils

Formation of present continents by breakup of Pangaea

Continued biological evolution

0

FIGURE 7-6

Geological and biological events on the same time scale.

some 4500 m yr ago and culminating some 1500 m yr later in the appearance of bacteria and blue-green algae. During these early eras protocells acquired and evolved a genetic and protein-synthesizing apparatus and, through this, an hereditable metabolism.

To place major events of both geological and biological significance in a time perspective, the two have been coordinated to a gross time scale in Figure 7-6. During a span of 1500 m yr, numerous protocell experiments could have been tested before unambiguous, truly biological forms certainly existed.

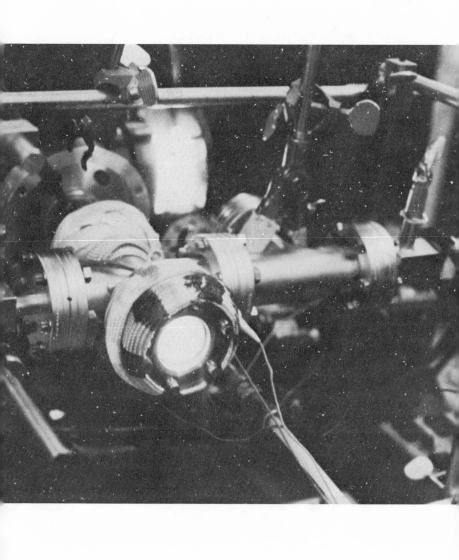

CHAPTER EIGHT

ORGANIC AUTOMATA

A lesser degree of complexity in an automaton can be compensated for by an appropriate increase of complexity of the instructions.

JOHN VON NEUMANN

Ultraviolet photolysis of methane. The reaction chamber shown here is stainless steel plated with gold to make its surfaces as chemically nonreactive as possible. Through various valves the chamber is evacuated and methane (or other gases) is introduced. A beam of ultraviolet radiation enters the interior of the chamber through a magnesium fluoride window, which passes all radiation of wavelength down to 170 nm. The source for ultraviolet radiation (not shown) is the giant argon plasma arc lamp at the NASA Ames Research Center laboratories. Since many of the products formed in these experiments are nonvolatile, they tend to condense on the surface of the magnesium fluoride window. For this reason the window is heated with a heat-tape about its periphery to keep these products in the vapor phase. (Photo courtesy of M. Dwyer, University of Pennsylvania.)

We have considered how the likely structures of protocells arose, but we have discussed neither their doings nor their potential for evolution. This chapter will speculate upon the design requirements and properties of the very simplest model organic automata and then relate the model system to likely events during the origin of the earliest protobionts.

An automaton is a thing regarded as capable of spontaneous action. Humans and bacterial cells are natural automata of marvelous complexity; artificial mechanical constructs such as computers and telephone switchboards are also automata, but of a far simpler nature. The thought experiment we shall attempt in this chapter will be to create the simplest possible organic automaton, using as boundary conditions observations from cell biology and our knowledge of the environment of 4000 m yr ago. We will see that such organic automata can be extremely simple, but only if the environment is complex.

Boundary Condition I: Biological Cell Theory

A phase boundary, or membrane, is required to isolate temporarily a portion of the environment. In aqueous systems a phase boundary must be hydrophobic. One can construct spherules of self-assembled lipids by casting lipids onto disturbed water surfaces. Indeed, Oparin's coacervates, Fox's proteinoid microspheres, and my organic microstructures all demonstrate that spheres self-assemble in aqueous systems from partly hydrophobic materials.

Boundary Condition II: Energy and Its Receptors

A vast number of possibilities seem to occur here, ranging from the use of sun-derived chemical potential energy locked in small organic molecules to the direct use of solar energy. I prefer the latter because it was and is by far the dominant form of energy available. The most energetic portion of the solar spectrum, then and for 2500 m yr to come, was short-wavelength ultraviolet radiation. This form of energy is intense enough to make and break chemical bonds between the bioelements directly, and it can easily be absorbed by a variety of organic compounds. Thus, for energy, let us use solar ultraviolet radiation.

Rather than wasting this energy by permitting it to play randomly over the spherules, let us use specific ultraviolet radiation receptors. The most obvious receptor is modeled after those used by photosynthesizing organisms—porphyrins. Porphyrins are planar conjugated systems built upon four pyrrole molecules held together by methylenic bridges. Various kinds of porphyrins are characterized by the specific groups substituted along the outer edges of the pyrrole groups (see Figure 8-1). All porphyrins have a number of useful properties in common. They all absorb both visible and ultraviolet radiation and thus serve as superb energy receptors. They are all hydrophobic: they are quite water insoluble, preferentially attaching and adsorbing to the hydrophobic phase boundaries of the organic automaton. And they all complex with metal ions. The four nitrogen atoms of the pyrrole rings all point inward; their spacing and charge are just right to hold metal ions. Metal ions thus held can undergo valence changes, permitting this organometallic receptor to move electrons from one state to another. Although porphyrins

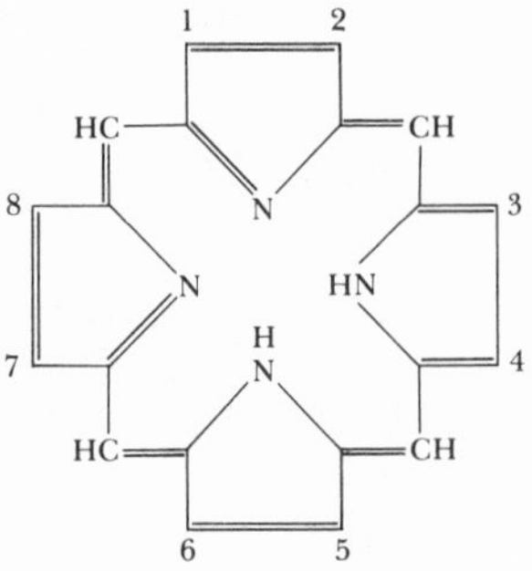

FIGURE 8-1

*The skeleton of a porphyrin molecule.
Specific porphyrins differ in having various
organic groups attached onto positions 1
through 8 and in complexing with a
metallic ion at the junction of the four
pyrrole nitrogens.*

are complex molecules, they can be formed from combinations
of pyrroles, which are easily made in chemical evolution simulation experiments.

Boundary Condition III: A Stable Environment

The complexity of biological systems permits them to
adapt to and modify a changing environment. For the organic
automaton we must assume a completely stable environment.
This assumption permits us to push potential complexity of the
automaton into the instructions—the environment itself. Hence
we will give the environment these parameters:

1. A steady, unlimited supply at constant concentration of
all small organic molecules needed.

2. A constant flux of radiant energy and constant physical
features such as temperature, pH, and salt content.

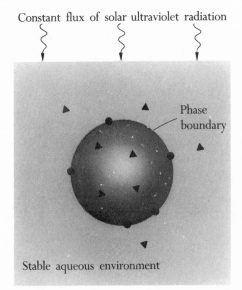

Constant flux of solar ultraviolet radiation

Phase boundary

Stable aqueous environment

▲ Small organic or inorganic target molecule

● Adsorbed porphyrin molecules

FIGURE 8-2

The simplest organic automaton. A constant flux of ultraviolet radiation impinges on phase-bounded receptor spheres that contain target molecules.

Since we are considering simple systems in order to observe general properties, we shall assume that membrane precursor molecules, porphyrins, and one other kind of small organic target molecule are present. (This latter molecule is representative of all molecules involved in biosynthesis.) Figure 8-2 summarizes a simple organic automaton.

What happens when we put all this together? Energy receptors—the porphyrins—absorb ultraviolet radiation. In this process, electrons of the receptors are raised in energy level; these become available to conduct reactions which must be driven, either directly or by creating a proton gradient. In the direct case we can imagine a small target molecule, say an amino acid, which must be activated so that it can then bind to another amino acid. In the proton gradient case we can suppose that high-energy electrons are pumped out of the spherule, inefficient as this mechanism may be. This circumstance sets up a charge gradient that enables chemical work—polymerization or activation of target molecules—to be performed.

Thus our small target molecules become activated and polymerized as a result of transformation of radiant energy by the receptors into high-energy electrons and then into chemical potential energy. Undoubtedly this will happen both inside and outside the spherule's surface. However, the volume enclosed is finite, whereas the exterior—since we consider it to be stable—is for all practical purposes infinite. The automaton will modify the nature of its interior space by using up most of its target molecules. As this occurs, a concentration gradient will arise. More target molecules will be present without than within, and as a consequence more such molecules will diffuse into the interior of the automaton.

Eventually each automaton will accumulate a high concentration of polymer, and reactions of target molecules will be inhibited. We might say that the spherule "dies"—it ceases to function. Continued energetic ultraviolet radiation will degrade its absorbed porphyrins and break down its membrane, releasing its compounds for reincorporation into succeeding generations.

Alternatively, high concentrations of internal polymer could interact with the phase boundary and with phase boundary precursors, causing an increase in phase boundary area and ultimately leading to fission of the automaton. The newly formed daughter automata would contain older portions of the parent as well as newly incorporated phase boundary and receptor materials. Note that in this growth alternative, the polymer synthesized by the parent becomes randomly distributed to the daughters.

Since by hypothesis our environment contains a stable

concentration of membrane and other precursors, we will observe a steady rate of synthesis of new automata by self-assembly, a fixed rate of synthesis of target molecules into polymer, and a constant rate of degradation of "older" automata.

A very small number of relatively simple organic molecules can mirror the activities of biological cells, given a complex, artificially stable environment. This is why our model required no enzymes or genetic apparatus. Does this model organic automaton provide any clues to the origin and direction of early cellular life? Let us expand upon the model automaton approach by considering how such a system might behave in the truly complex environment of the warm little pond.

In Chapter Seven we made the case that protocells were formed from polymeric material at the same time as the formation of small organic molecules. Such protocells present a complex phase boundary, which has a hydrophobic nature and will act selectively to adsorb porphyrins and similar small organic molecules. Porphyrins adsorbed to the protocell material can act as ultraviolet radiation receptors, set up proton gradients, and convert radiant energy to chemical bond potential energy.

Target molecules constitute a large set: amino acids, nucleic acid bases, sugar precursors, inorganic phosphate, and so on. From these, random polymers of amino acids (proteins), of nucleotides (nucleic acids) and of sugars (polysaccharides) can be formed. Recall that the organic soup is most dilute. Any one amino acid will be present at a concentration of 10^{-7} M; this means that about 3×10^4 molecules of each amino acid will occur in each 500 μm^3 (the volume of an average protocell of radius 5 μm). Assuming 200 different kinds of small molecules, we guess that every newly formed protocell could enclose a total of 6×10^6 total small organic molecules. The enclosed volume is shielded and is small relative to the exterior. As target molecules react within the protocell to form random polymers, a concentration gradient for each one arises, and more target molecules diffuse into the protocell.

In this way the protocell overcomes the concentration gap, although its efficiency does not compare with that of evolved biological systems. Complexity—as expressed in possibilities—is increased enormously in this first generation of protocells over our automaton model, because of the large number of reactions and interactions possible within the large set of target molecules. After the first generation, however, complexity increases at an even greater rate because random polymers of amino acids and nucleic acids influence the course of reactions among the target molecules.

Gary Steinman has shown that random polymers of amino acids are not "statistically" random. Instead, the nature of the "tail" of each amino acid sways the purely statistical likelihood that might otherwise be expected. Large bulky "tails" tend not to be adjacent; similarly charged "tails" tend not to be adjacent; and so on. Thus abiological amino acid polymers are not purely statistically random but reflect the individual molecular architecture of their constituent amino acids. This might be true for any abiological polymerization process that uses more than one kind of small precursor molecule.

As Sidney Fox has shown, randomly formed proteinoids possess a wide variety of low-level catalytic abilities. These abilities introduce another dimension of complexity. Protocells containing randomly made protein possess a wide spectrum of faint catalytic ability, and through this catalytic haze they further extend their metabolic potentials.

As a protocell accumulates various polymers within, it can also add onto its membranous surface. As surface area increases, the weak interactive forces holding the membrane together become insufficient; the protocell must bud or split in two. These hypothetical protocells possess no genetic material or specific protein-synthesizing apparatus. Upon fission, whatever of value

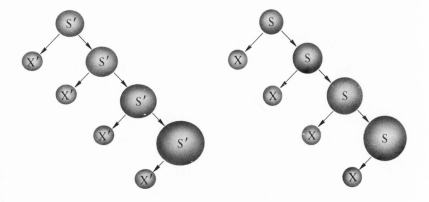

FIGURE 8-3

Accumulation of effective lines of protocell descent. A single protocell, "S," possessing synthetic abilities, can at best form a single line of descent, since it has no genetic mechanism to distribute copies of these functions to daughter protocells. Daughter buds labeled "X" did not receive synthetic functions and can no longer bud.

is contained within the parent protocell is by chance retained by one daughter and lost to the other. However, a potential line of descent is already discernible. Those protocells that are more effective at constructing metabolically useful proteins will form unilinear descent lines that can be positively selected for even at this early stage. This is an inefficient and passive form of selection. Of the entire ensemble of protocells, those lines of descent that are successful at making random polymers will accumulate. Figure 8-3 elaborates on these possibilities.

At this early stage in the origin of life, we are thus presented with an ultraviolet photosynthetic heterotroph. The protocell uses ultraviolet radiant energy to create polymers; it requires a continuous but dilute steady supply of all amino acids, nucleic acid bases, sugars, and other small organic molecules. Random polymers of amino acids complexed with metallic ions endow the protocell with a great range of low-level catalytic abilities which are inherited by single lines of descent, as are random polymers of nucleic acids and sugars.

Protocell descent lines such as these undoubtedly flickered into existence only to be lost by the drying of pools or to sink to such depths that radiant energy was no longer available. These early forms could, however, self-assemble rapidly and repeatedly. One can imagine pond after pond serving as foci for innumerable protocell experiments over the course of 500 to 1000 m yr. Although an ultraviolet photosynthetic heterotroph seems the most likely initial experiment, many potential environmental niches come into being once such a protocreature exists. Consider a protocell descent line, originally ultraviolet photosynthetic, but containing a random protein set that will effect continued amino acid polymerization using the chemical potential energy in the detritus of degraded protocells. Even if it sinks out of sight of radiant energy, this protocell can still continue to function, perhaps to increase the chances of alternate descent lines arising in this new heterotrophic niche.

In developing the general argument for the ultraviolet photosynthetic heterotroph and for nongenetic lines of descent, we have tacitly assumed that polymers of sugars, amino acids, and nucleotides were formed directly by energetic collisions of activated small precursors. This assumption needs revision; we must now deal with the mechanism of protocell polymer synthesis.

Most biological polymerizations take place by the joining of two molecules with the apparent elimination of a molecule of water. These reactions are quite mistakenly called anhydrous condensation reactions because the overall equation looks like this:

$$
\underset{\text{amino acid}}{
\begin{array}{c}
\text{R}-\text{C}-\text{C} \overset{\displaystyle\text{O}}{\underset{\displaystyle\text{OH}}{\diagup}} \\
\text{N}\quad \\
\diagup\; \backslash \\
\text{H}\;\;\text{H}
\end{array}}
\;+\;
\underset{\text{amino acid}}{
\begin{array}{c}
\text{R}-\text{C}-\text{C} \overset{\displaystyle\text{O}}{\underset{\displaystyle\text{OH}}{\diagup}} \\
\text{N}\quad \\
\diagup\; \backslash \\
\text{H}\;\;\text{H}
\end{array}}
\;\longrightarrow\;
\underset{\text{dipeptide}}{
\begin{array}{c}
\text{R}-\text{C}-\text{C}-\text{N}-\text{C}-\text{R}' \\
\text{N}\qquad\text{H} \\
\diagup\; \backslash \\
\text{H}\;\;\text{H}
\end{array}}
\;+\; \underset{\text{water}}{\text{HOH}}
$$

To join two molecules of amino acids we must eliminate a molecule of water. Now the problem is that in our primeval pond amino acids, or other precursors, occur in exceedingly dilute aqueous solution. Water is in such gross excess that this or any other anhydrous condensation reaction will barely proceed. In principle this difficulty seems to apply also to polynucleotide synthesis and polysaccharide synthesis, but we shall consider polypeptide synthesis as a general example.

Three ways are known to overcome the water problem: water can be removed by drying; amino acids can be made chemically reactive; and enzymes can force activated amino acids together.

The difficulty with the drying hypothesis is that relative amino acid concentrations are quite low even within protocell interiors, although there is no doubt that large numbers of protocells were frequently exposed to alternate drying and wetting cycles. Nevertheless the possibility is real that some protocell polymer synthesis could occur by this mechanism.

The enzyme hypothesis presents a twofold difficulty. First, at this early time enzymatic activity took place only at a very low level, mediated by randomly synthesized proteins. Second, all known enzymes do not join amino acids as such but convert them first to reactive forms which then are condensed.

The most general way to overcome our difficulty—and the only reasonable way left open to us—is to make amino acids chemically reactive. Throughout the ecosystem, the principal mechanism for making small organic molecules in water solutions chemically reactive is to couple these molecules to some form of phosphate.

Transfer of the phosphate group releases or absorbs energy; such transfers are the medium through which all biological energy is stored or used to effect condensation or metabolic reactions. Energetic bonds formed by various phosphates with organic compounds drive all biological reactions today, and this could have been the case in the world of protocells too.

We will consider a simplified example of the use of phosphate-activated condensations and of one probable route by which energetic phosphates could have been made.

Consider this compound, called pyrophosphate:

$$HO-\overset{\overset{\textstyle O}{\|}}{\underset{\underset{\textstyle OH}{|}}{P}}-O-\overset{\overset{\textstyle O}{\|}}{\underset{\underset{\textstyle OH}{|}}{P}}-OH$$

which we will now write as P—P. The phosphorus-to-phosphorus bond contains potential chemical energy which can be released to activate organic compounds. For example, glycine, the simplest amino acid, might be activated as follows:

| glycine | pyrophosphate | phosphoric acid ester of glycine | inorganic phosphate |

Some energy is lost in the phosphate transfer to glycine, but the ester bond of phosphate to glycine contains more than enough energy to drive a coupling reaction:

glycine-P + glycine-P ⟶ glycine-glycine-P + P_i

two activated glycine molecules   diglycine   inorganic phosphate

Notice that the diglycine in this example remains activated. Notice also that we no longer face the problem of anhydrous condensation reactions! Phosphate activation and transfer avoids the difficulty. Almost any small organic biomolecule can easily enter almost any reaction in the presence of large amounts of water if and only if it is phosphate-activated. The general conclusion is:

> *Protocell synthesis of polymers proceeded by phosphate-activated intermediates.*

Anhydrous condensation reactions are not a feature of our present biochemistry and were not in protocell times either, if we are a reflection of those times. Phosphate transfer reactions are now and were then the only important way to effect condensations in a water-dominated environment.

One serious consideration remains. To drive phosphate transfer reactions requires a source of high-energy phosphate —pyrophosphate (P—P) is the simplest form. This molecule is unstable in aqueous solutions, and all a protocell might encounter would be dissolved inorganic phosphate, P_i. Where could the pyrophosphate needed for all the protocell reactions come from?

When a bounded system with energy receptors is exposed to radiant energy, electrons are moved to higher energy levels; they can literally be pumped across the membrane. This sets up a charge gradient that can be intense enough to drive the following reaction:

P_i + P_i ⟶   P—P

inorganic phosphate   pyrophosphate

The photosynthesis of pyrophosphate was probably an important feature of the primary metabolism of protocells. As one of many possible points of comparison, present-day photosynthetic cells set up such proton charge gradients to drive the synthesis of adenosine triphosphate from adenosine diphosphate, either directly—as in cyclic photophosphorylation—or indirectly, as in noncyclic photophosphorylation. This process is vastly more efficient than the pyrophosphate-formation mechanism proposed for protocells, but it is basically the same scheme.

The substitution of phosphate transfer reactions for organic anhydrous condensation reactions, which forms the basis of our biochemistry, began with the first protocell. Perhaps it is for this reason that the phosphorus cycle drives all other bioelemental cycles in coupled reactions.

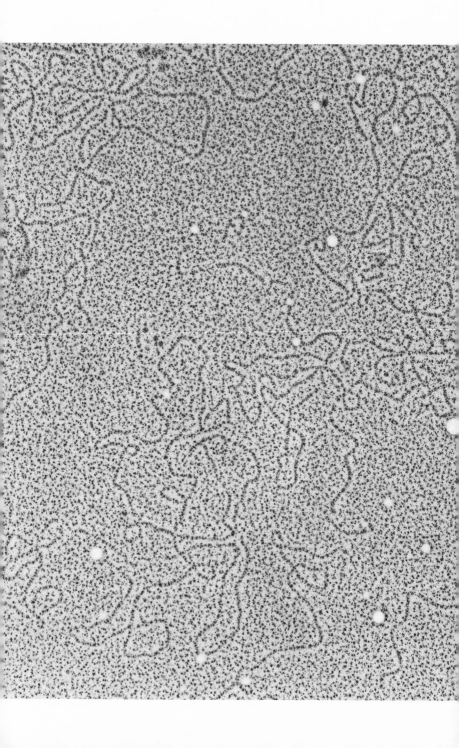

CHAPTER NINE

GENETICS

Einstein was right, God does not play dice with the Universe.

This is an electron photomicrograph of rat liver DNA and rat liver mitochondrial DNA. The magnification as seen is 72,000 ×. DNA molecules have been spread out upon a very thin colloidin membrane and shadowed with platinum-palladium. The white spots are minute holes in the supporting membrane. DNA molecules, which are about 20 Angstroms (1 Angstrom = 1 × 10^{-10} meter) in width, are seen as darker, randomly coiled and dispersed strands. Mitochondrial DNA is distinguished from nuclear DNA in that the former are coils that topographically bind back toward each other—that is, they are circles of DNA. Nuclear DNA is seen here as linear segments. (Photo courtesy of Dr. D. E. Philpott, NASA Ames Research Center.)

A plausible model can be drawn for protocells arising very early in unstable shallow ponds. But these protocells are far removed from anything we would call a cell, since they contain no genetic apparatus and no specific protein-synthesizing system. Any random protocell polymer that was formed could at best echo down a short corridor of one-lined descent, ultimately to be degraded. Our object in the next few chapters will be to show how a functioning, though simple, genetic apparatus and a specific protein-synthesizing machinery could arise within protocells. The essence of our approach will be to consider carefully the interaction potentials of all those polymers randomly formed within protocells.

As background, this chapter surveys the essentials of the current biological genetic and protein-synthesizing apparatus. We will use word "information" to mean specific amino acid or nucleic acid base sequence, and no more.

The genetic apparatus is the deoxyribonucleic acid (DNA) molecule in which information is stored. Components of this two-stranded polymer are the nucleic-acid bases adenine (A), thymine (T), cytosine (C), and guanine (G), each of which is attached to a sugar (deoxyribose). Successive sugars are held together by phosphodiester bonds. Figure 9-1 presents a two-dimensional view of this aspect of the molecule. The backbones of the molecule are the two repeating sugar-phosphate chains. The information of the molecule resides in its base sequence. Each strand of DNA consists of a sugar-phosphate backbone with a base projecting from every sugar. Each of the two strands has a chemical direction, and a complete molecule of DNA is in fact a

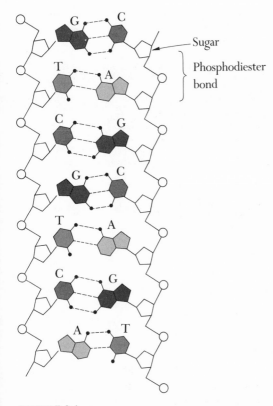

FIGURE 9-1

A two-dimensional representation of a DNA duplex molecule. (After Arthur Kornberg, "The Synthesis of DNA." Copyright © 1968 by Scientific American, Inc. All rights reserved.)

complex mode of the two strands coiled about each other, running in opposite directions and held together by weak hydrogen bonds formed between the bases (dotted lines). A DNA molecule contains hundreds of thousands of base pairs. The two strands of the molecule can be held together by hydrogen bonds only if they are complementary, that is, if they match. A can form two hydrogen bonds only with T; G can form three hydrogen bonds only with C. These *base pairing rules* are fundamental to both the copying and the functioning roles of the genetic apparatus.

If we use the following base sequence shorthand for one piece of a strand:

A G C T A G C T A G C T

then by the base pairing rules, the complementary strand must be:

T C G A T C G A T C G A

and the duplex molecule would be the complementary pair:

A G C T A G C T A G C T
T C G A T C G A T C G A

The molecule would also be coiled about itself.

The base pairing specificities give the DNA molecule a potential for information transfer. As the duplex molecule replicates, it forms two almost identical daughters. The daughters are never exactly identical because of errors that can occur in the base pairing process during replication.

When a large molecule of DNA is to be replicated in a growing cell, a great number of events occur. First, a large "unwinding" protein opens the duplex molecule at a specific growth site so that the bases within become exposed:

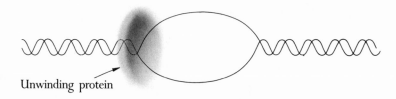

Unwinding protein

Then another large polymeric protein complex occupies sites on the single strands:

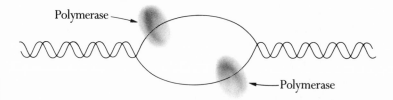

Polymerase

Polymerase

The polymerase proteins then accept and test precursor molecules from a pool of A, G, C, and T nucleotides until the base pairing rules are satisfied. When a successful match of new to parental base is made, the polymerase protein jumps one base notch to repeat the matching process and to join two adjacent daughter bases together.

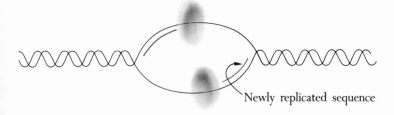

Newly replicated sequence

In this way the parental base sequence duplex can be copied almost exactly, creating two daughter base sequence duplexes. A conspicuous feature of DNA replication is that its fidelity is enhanced by the presence of replication enzymes that can discriminate between matches and mismatches of base pairs. Accurate replication is a consequence both of the structure of the DNA molecule and of the nature of its replicating enzymes.

How can base sequencing information be translated into the sequence of amino acids in a protein molecule? We will see that this process, like DNA replication, results from both the nature of the base pairing rules and the interaction of quite complex proteins.

Ribonucleic acids (RNA) are single-stranded copies of DNA base sequences. The basic structure of RNA, like that of DNA, consists of a sugar-phosphate backbone. Attached to each ribose sugar is one of the bases A, G, C, or U (uracil, a close relative of thymine which has the same base pairing configuration).

Synthesis of all RNA molecules takes place through the mediation of a set of polymerase proteins; the RNA strands are base-sequence-complementary copies of specific stretches of DNA.

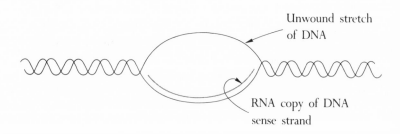

Unwound stretch of DNA

RNA copy of DNA sense strand

A large RNA polymerase enzyme, with its complex of coenzymes, begins its copying activity at a specific DNA base sequence site and matches complementary RNA bases to DNA bases according to the base pairing rules. When a match is made, adjacent RNA bases are polymerized by another site on the polymerase enzyme; the process is repeated down one strand of DNA until the polymerase enzyme encounters a specific DNA base sequence "stop" signal. The single-stranded RNA transcripts then fall off the DNA molecule and become available for protein synthesis. Only one of the two DNA strands is copied in each instance; the RNA polymerase enzyme recognizes which DNA strand is to be copied.

By this one mechanism, three classes of RNA transcripts are produced from DNA: transfer RNA, messenger RNA, and ribosomal RNA.

Ribosomal RNA is enzymatically cut, modified, and wrapped with about 50 different protein subunits to form a ribosome. This structure, a universal biological organelle, is the site of actual protein synthesis. Essentially its role is to provide a complex, coordinated surface to bring all reactants together in proper spatial relationships.

Messenger RNA carries in its base sequence the code for a linear sequence of amino acids. It remains as a linear copy of DNA sequences and finds its way onto a ribosomal surface, where the decoding is accomplished by yet another class of RNA.

There are more than 20 different base sequence classes of transfer RNA molecules. These molecules are all synthesized as linear transcripts of DNA, but they are enzymatically modified by a series of enzymes: some bases are methylated, others are hydroxymethylated, still others are reduced, and so on. During the modification process all transfer RNA molecules, which originally were linear, bond back upon themselves to form cloverleaf-shaped molecules. At one free end of the molecule a common triplet of bases (CCA) is always added enzymatically. This end becomes the site of amino acid attachment. At the top

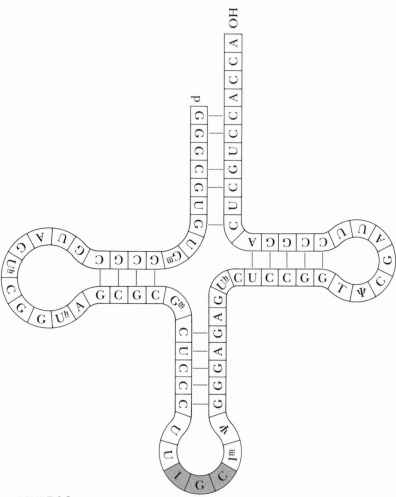

FIGURE 9-2

Base sequence and structure of a typical transfer RNA molecule. (After Robert W. Holley, "The Nucleotide Sequence of a Nucleic Acid." Copyright © 1966 by Scientific American, Inc. All rights reserved.)

of the middle "leaf," three bases will protrude. Figure 9-2 shows the structure of a typical transfer RNA molecule. Because of the geometry of the ribose-phosphate backbone, three or more bases must always protrude from the "leaf" end of the molecule.

Each class of transfer RNA becomes bonded to one specific kind of amino acid by its own specific enzyme. Each of these enzymes has sites that recognize sequences on its own transfer RNA molecules, sites that recognize one particular amino acid, and sites that bond the amino acid onto the adenosine of the common ACC terminus.

Cells that are engaged in protein synthesis are packed with ribosomes, contain pools of charged transfer RNAs, and actively synthesize messenger RNA transcripts.

Figure 9-3 gives a simplified diagram of protein synthesis. A messenger RNA bonds weakly to one end of a ribosome. A special "start" complex, consisting of transfer RNA and methionine (formylmethionine in bacteria), always begins the decoding process. In this process, the three projecting bases (the anticodon) of the transfer RNA find a matching triplet (codon) on the messenger and bond weakly to it according to the base pairing rules. The second set of three messenger RNA codon bases, in turn, can accept only a specific anticodon match for the next transfer RNA—amino acid complex; all other, nonmatching anticodons are rejected. The two decoded amino acids are now in close juxtaposition within two grooves on the ribosome surface, and a peptide bond between them is formed enzymatically. With some of the energy released from peptide bond formation, the messenger RNA jumps forward one triplet of bases; the process of transfer RNA—amino acid anticodon-codon selection and subsequent peptide bond formation is repeated again and again until a "stop" triplet, which codes for no amino acid, is reached.

Complex as all this seems, it is an oversimplification: many more layers of intricacy are in fact involved in all phases of the process. For our purposes it is sufficient to recognize that the entire genetic and protein-synthesizing apparatus uses a large array of multisite proteins and cofactors, each bearing many specific sites that recognize both nucleic-acid base sequences and amino acids of various kinds. Base sequence information of

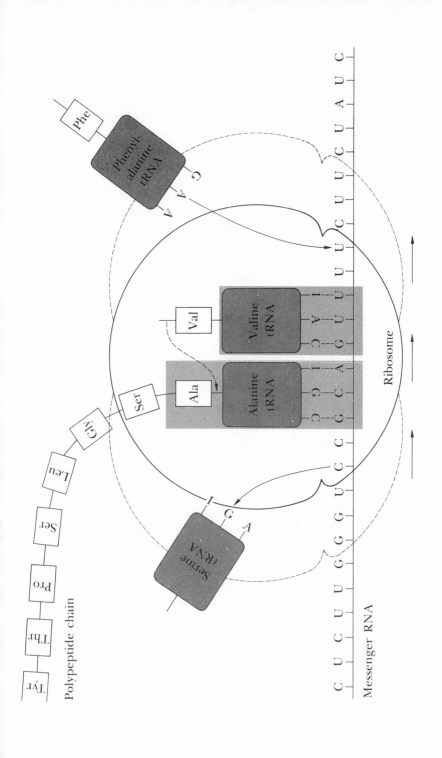

Polypeptide chain

Ribosome

Messenger RNA

DNA is transcribed into shorter sequences of various RNAs, which are modified by complex proteins. A family of specific proteins matches the correct transfer RNA to the right amino acid. Base sequence is translated stepwise into amino acid sequence, so that each triplet of DNA bases ultimately corresponds to a single specific amino acid. The genetic code seems to be universal: a triplet that codes for a given amino acid in those bacteria that have been studied will also code for that same amino acid in plant or animal cell preparations.

FIGURE 9-3

A generalized overview of protein synthesis. (From F. H. C. Crick, "The Genetic Code: III." Copyright © 1966 by Scientific American, Inc. All rights reserved.)

CHAPTER TEN

ORIGIN OF THE GENETIC PROTOCELL

We all agree that your theory is mad.
The problem which divides us is this:
is it sufficiently crazy to be right?

NIELS BOHR

A view of California from an orbit about Earth. (Courtesy of NASA.)

*I*t is far from obvious how the complex system described in Chapter Nine, involving such an interrelationship between protein and nucleic acid, could have evolved from protocells as simple as the ones we have postulated. In this chapter we shall outline one possibility, attempting not to delineate the entire evolution of this system (which probably required more than 1000 m yr to evolve), but only to show how its most fundamental features might have arisen.

Protocells all faced a common problem. Even though by chance a protocell might contain a useful protein, there was no possible way to assure that copies of this protein could be made to be passed to daughter protocells. Hence all protocell lines of descent were one-sided. There was no reproduction to excess of specific hereditable features that could be subjected to natural selection.

Protocells did contain randomly made polymers, but let us not assume that because biopolymers are large, primitive polymers were also. It is more plausible that primitive polymers were quite small. Primitive "proteins" were about 5 to 7 amino acid residues long, compared to 100 to 300 residues for present-day proteins. Polynucleotides were 10 to 12 bases long rather than millions of bases. Many experiments dealing with polymerizations of amino acids and nucleic acids under primeval conditions support these general statements. If primeval polymers were so small, how did they function? The active catalytic site of an enzyme is almost always a small number of amino acid residues. The balance of the large biomolecule is involved with control, stabilization, specific site attainment within the cell, and with a host of other properties that are vital for integrative control but not for the catalysis of a specific reaction.

We can accordingly reduce the enormity of the problem somewhat by restricting our initial requirements only to the synthesis of small specific peptides and oligonucleotides (short polymers of nucleotides).

In like fashion we will not invoke highly evolved enzymes or special clay surfaces. To explain the origin of the genetic and protein synthesizing system at the most primitive level, we need only consider the interactions of two kinds of short polynucleotides to make a short peptide.

To begin, we will assume the presence of a small prototype *generator* RNA, which fulfills both the genetic and template roles, and the existence of a small family of prototransfer RNAs. How might this system work? Let us separate the genetic aspect of the primitive system from the protein synthesizing aspect and approach each in turn.

Imagine a protocell that contains a group of polynucleotides some 12 units long (a dodecamer). In some cases this might be a linear molecule, but, since precursors are dilute and infrequent, the most probable occurrence (this can be demonstrated in the laboratory) is for the dodecamer to join head-to-tail to form a small circular molecule. Circular forms of polynucleotides are chemically more stable than linear forms, since no free hydroxyl or phosphate groups are exposed to random degradation. The bases, exposed on the circle, are free to act as a template for pairing of free nucleotides according to the base pairing rules. Slowly, and with some mistakes, a complementary polymer can be copied without enzymes from this circular generator template molecule. However, as in some bacterial systems found today, the copy complement can continue past the starting point to yield a longer molecule. Replication can start at any point at random. As a consequence:

> Sequences of the complements will be circular permutations of the generator sequence.

In this way small, closed circular generators can form complementary copies of themselves, as well as both small and large linear molecules. Figure 10-1 diagrams these aspects of primitive replication. The linear molecules are random permutations of the repeated 12-unit basic circular sequence, since copying could start from any position on the small 12-unit generator.

Let us assume that the role of circular complements corresponds to that of messenger RNA and that the role of circular generators is genetic. The circularly permuted small fragments (later the larger ones) will serve as a set from which primitive transfer RNAs can be selected.

At this point a major problem arises. The anticodons of present-day transfer RNA are matched to the correct amino acids by large, highly evolved proteins. How can primitive systems attain a correspondence of specific matching without these enzymes? We will discuss some guesses below, aspects of which border upon violating our principle of not admitting into the argument the likelihood of historical accident.

Before coping with this difficulty, let us simplify some of its aspects. Although 20-odd different amino acids are now used in proteins, they fall into four groups: the basic, such as lysine, which contain more amino groups than carboxyl groups; the acidic, such as aspartic acid, with more carboxyl groups than amino groups; the aliphatic, such as leucine, with only a carbon skeleton and one each of amino and carboxyl groups; and the

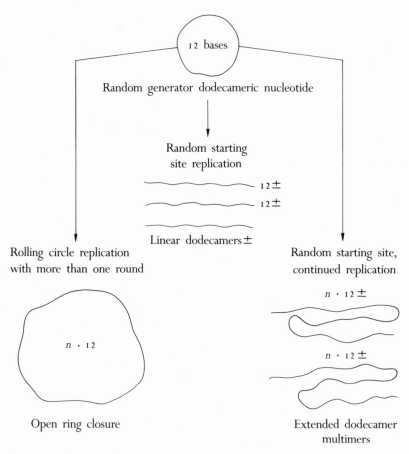

FIGURE 10-1

Possible replicates that could be formed from a primitive dodecanucleotide.

polar, such as serine, which have additional hydroxyl groups. A small, primitive peptide manifesting some catalytic activity could tolerate any of a variety of different amino acids *of the same class* at a particular position within the peptide without too great a loss of function. Our task, then, is to find plausible ways to couple one of four classes, rather than one of twenty specific amino acids, onto specific primitive transfer RNA molecules.

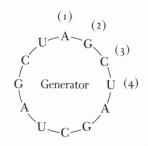

Short, linearly permuted
complementary copies at site

(1)	(2)	(3)	(4)	
U	C	G	A	
C	G	A	U	Amino acid
G	A	U	C	discriminator sites
A	U	C	G	
U	C	G	A	
C	G	A	U	Anticodons
G	A	U	C	
A	U	C	G	

FIGURE 10-2

Base sequences copied from various points on the generator lead to the formation of short, linearly permuted, primitive transfer RNA molecules in which discriminator sequences bear definite relationships to anticodon sites.

One proposal is that a short stretch of base sequences near the open end (the adenylic acid end) of primitive transfer RNAs acted as a discriminator; Figure 10-2 illustrates this. A certain sequence of discriminator bases might select for one class of amino acids over others by charge, hydrogen bonding, or other weak associative or repulsive forces. In this fashion, families of primitive transfer RNAs having various discriminator sequences would come to be associated with specific functional groups of amino acids. Since we have earlier assumed that the amino acids are activated as the phosphate esters, those that can approach —or not be repelled from—the discriminator sites would be able to bond to the terminal group of the transfer RNA, yielding an activated, specific primitive transfer RNA. This bonding of amino acid to RNA could be accomplished with pyrophosphate bond energy.

This scheme seems to present one great difficulty: it is not immediately apparent in what way any correspondence might exist between the anticodon and discriminator ends of the primitive transfer RNA.

But a definite correspondence does exist and can be demonstrated! Imagine a small dodecanucleotide generator ring sequence. From this sequence we make linear copies, starting at any point at random. All copies are to be short and of the same average size. We will call one end of the linear copies the discriminator region; the other region, a fixed distance away, is the anticodon. Since all linear copies are *permutations* of the generator sequence, the invariant result is:

Unique anticodons are specified for every discriminator region in any permuted transcript.

Figure 10-2 gives several examples for a generator sequence selected at random. Try others and verify that their anticodons are unique.

The model of short primitive transfer RNA just depicted illustrates the principle of permutations of one-dimensional sequences; it may indeed mirror the first transfer RNAs. Whether we copy a short or a long generator molecule to form linearly permuted copies, the conclusion is inescapable that a correspondence must exist between discriminators and anticodons.

Larger transfer RNA molecules are no doubt a product of later evolution, but the permutation rule still applies. Discriminator and anticodon can only be related by permutation. Manifred Eigen devised a very simple and revealing game: one throws a four-faced die, with the faces labeled A, C, G, and U, for 80 successive times and records the random "base" sequence generated by that set of throws. This represents one random polynucleotide. The object of the game is to use this sequence and the base pairing rules to construct as many consecutive back-bonded base pairs as possible within the molecule. Almost always a cloverleaf-shaped structure reminiscent of transfer RNA results! This seems to indicate that random linear RNA molecules frequently seek out a gross structural aspect like that of transfer RNA as they assemble themselves into more stable structures. However, correspondence between the amino acid bonding end and the anticodon remains in the domain of permuted sequences from a single generator. There is no correspondence of discriminator and anticodon sites in a whole family of *randomly* generated sequences such as Eigen's, since they are not descended from a common generator.

How good is the evidence that sequences of bases at the discriminator site will actually interact specifically with the various amino acid classes? Qualitatively the evidence is quite good. Carl Saxinger and Cyril Ponnamperuma have shown clearly that specific amino acid–oligonucleotide interactions do occur. To support this statement we can also present the contemporary observation that enzymes (polymers of amino acids) can and do recognize specific short nucleotide sequences quite accurately.

Another potentially disturbing problem arises with this or any other explanation of the origin of primitive transfer RNAs. No compelling reason seems to require us to have the genetic code we do and not some other permutation of it. Figure 10-3 displays the present genetic code. Composed of 64 (4 times 4 times 4) base triplets coding for 20 amino acids, the code is degenerate; that is, more than one triplet codon specifies a single amino acid. But why this code and not some other? Why do UUU and UUC code for phenylalanine in all organisms? One is sorely tempted to invoke historical accident and natural selection and to imagine possible worlds of equally likely codes. But this is against the rules of our game of scientific inquiry! Let us consider some other answers to this question.

The rule of a common trophic level of ecology for all organisms requires only that all organisms have a similar small molecule biochemistry; it does not appear to require a universal

First base \ Second base	U	C	A	G	Third base
U	UUU, UUC: Phe UUA, UUG: Leu	UCU, UCC, UCA, UCG: Ser	UAU, UAC: Tyr UAA non UAG non	UGU, UGC: Cys UGA non UGG Trp	U C A G
C	CUU, CUC, CUA, CUG: Leu	CCU, CCC, CCA, CCG: Pro	CAU, CAC: His CAA, CAG: Gln	CGU, CGC, CGA, CGG: Arg	U C A G
A	AUU, AUC, AUA: Ile AUG Met	ACU, ACC, ACA, ACG: Thr	AAU, AAC: Asn AAA, AAG: Lys	AGU, AGC: Ser AGA, AGG: Arg	U C A G
G	GUU, GUC, GUA, GUG: Val	GCU, GCC, GCA, GCG: Ala	GAU, GAC: Asp GAA, GAG: Glu	GGU, GGC, GGA, GGG: Gly	U C A G

FIGURE 10-3

The genetic code. Depicted are messenger RNA triplet codons and their assignments to amino acids. Codons labeled "non" do not specify any amino acid, but instead cause termination of protein synthesis.

genetic code. The set of bases—A, G, C, and U (or T)—is restricted, but the code itself is not fixed.

Using the arguments of Darwinian natural selection, one can make a case for universality of the code in all complex nucleated organisms. These cell types are invariably the result of combinations and fusions of various symbiotic procaryotes. Vestiges of these conjoint fusings remain with us today. Chloroplasts of green plants, for example, are eucaryotic organelles that possess their own limited protein-synthesizing and genetic apparatus. Mitochondria of all eucaryotic cells are similar in this regard. The universality of the code is an important requirement for these symbioses, since the cell proper supplies all transfer RNA molecules both for itself and for its once symbiotic associates.

During the early evolution of the procaryotes, no such force would have required a universal code. If—and this is a big if—no underlying chemical or physical reason requires a universal code, then we might expect to find hints of alternative codes in various long-separated branches of the bacteria.

The evidence now available that supports the statement that the genetic code is universal is based on a single procaryote, *Escherichia coli,* which has a code similar to that of the eucaryotes. As yet no other evolutionarily far removed procaryotes have been studied to determine whether the genetic code is indeed universal for all procaryotes.

We thus are left with the conclusion that we have insufficient evidence to state unambiguously that the code is universal. Perhaps a survey of bacteria such as the Clostridia will one day aid us in this endeavor. More exciting and to the point—and unlikely in our time—would be a study of the code of an extraterrestrial organism.

Leaving unresolved the question of universality, let us try to reconstruct how a primitive "genetic" protobiont might synthesize a specific protein and how it might pass this ability to its daughters.

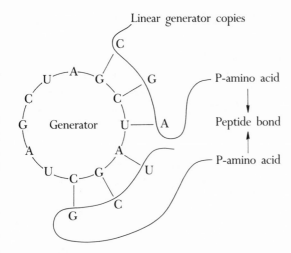

FIGURE 10-4

A model for primitive peptide synthesis. The short, circular generator sequence serves both as the genetic material and as the message for peptide synthesis. Linearly permuted copies of the generator sequence, which have amino acids of a specific class bound as the phosphate ester to the discriminator site, align themselves according to the base pairing rules to the generator (messenger), permitting juxtaposition of specific amino acid classes. Since these amino acids are activated as the phosphate esters, the energy required to form the peptide bond is available.

According to our hypothesis, a protocell contains a circular generator and a family of short (later longer) primitive transfer RNA molecules, each of which is charged with amino acids representative of its proper discriminator class. The object is to synthesize a small five- to seven-unit peptide of specific sequence (see Figure 10-4).

Assume that the circular generator, or copied complements

of it, serves both as the primeval genetic material and as the counterpart of messenger RNA. To obtain sequence-specific peptides, we need only conceive of amino acid—charged primitive transfer RNA molecules interacting with the messenger so that appropriate representatives of the amino acid classes are juxtaposed correctly to form peptide bonds. In many instances anticodon—codon sets will be out of register, slowing the process. But sometimes the process will work. Thus a genetic protocell can enter the brave new world of Darwinian natural selection, since it now can make copies of specific peptides and of its primitive genetic material. All else is a matter of history. Or is it?

CHAPTER ELEVEN

EVOLUTION OF GENETIC PROTOCELLS

Chemistry is good, Nature is better.

EFRAIM RACKER, *Mechanisms in Bioenergetics*

Fossil trilobites. These trilobites, along with brachiopods and sponges, provide fossil evidence of the complex multicellular life that arose about 600 m yr ago. Some 2000 m yr of microbial and unicellular evolution preceded these forms. Shown are several clustered examples of Ellipsocephalus hoffi. *(Photo courtesy of Tim Hall, Foothill College, Los Altos, California.)*

Once a crude genetic apparatus appeared, those lines of protocells that possessed it could pass on to both daughter cells the ability to make small specific polypeptides. Lines of descent then provide families of related genetic protocells. These groups now become subject to true Darwinian natural selection. We will consider this mechanism briefly in speculating upon the further development of early cellular life.

The crude genetic protocell contained only a few short, circular generator nucleic acids and a few short, linearly permuted primitive transfer RNA nucleic acids. This inventory is barely sufficient to make one or two polypeptides each four or five residues in length. How can this small primitive system evolve the capacity to make more specific polypeptides and thereby gain in metabolic functions?

First, circular generators of, say, 12 nucleotides can be replicated continuously, resulting in much larger molecules which themselves can circularize. Figure 11-1 depicts such a process. For this to occur we might assume the presence of some peptides that catalyze oligonucleotide formation.

This process creates two groups of new, larger molecules: large circular molecules and large permuted linear molecules. The former can act as new generators, since they possess the original primitive generator sequence duplicated many times.

Primitive generator

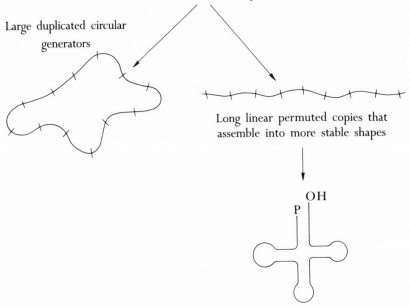

G—C—U—A—G—C—U—A—G—C—U—A—G ···

Results from continuous replication

Large duplicated circular
generators

Long linear permuted copies that
assemble into more stable shapes

P | OH

FIGURE 11-1

Replication of circular generators to make larger and more complex structures.

The large linear molecules will self-assemble to the more stable cloverleaf conformation by back-bonding. These molecules constitute a family of permuted, duplicated original generator sequences which can serve as larger primitive transfer RNAs. Thus a genetic apparatus of 80 or more nucleotides is available, permitting synthesis of repeating polypeptides as large as 25 residues, of several smaller peptides, or of peptides of intermediate size.

Copying primitive or intermediate generators by base pairing, using primitive peptides, is certain to be an inexact process. Large generators will contain the primitive generator sequence duplicated many times; some of these duplications will contain deviate bases. Hence considerable variation—mutation—of the primitive genetic material occurred, and natural selection can begin to take place.

Genetic protocells could quite rapidly have developed, by duplication and reduplication, the ability to synthesize large proteins having many and varied functions. The generator itself, interacting with large proteins, could thus have evolved into a large circular protochromosome. With large molecules having functional abilities, we have a primitive but unambiguously biological cell.

We have assumed that during this time the environment provided all necessary small molecules in dilute but essentially continuous supply, and that chemical energy was available as a result of the photosynthetic use of solar ultraviolet radiation to make pyrophosphate. When this environment becomes populated with primitive biological cells, it will change. Some small molecule nutrients will be used up faster than the environment can supply them. Selection pressures in favor of those cells that can modify a related available compound will be enormous. Using this reasoning, Norman Horowitz has visualized the development of cellular metabolism as a step-by-step *backward* selection, beginning with the more complex small molecules and proceeding to the simpler ones.

All metabolism is a series of enzyme-mediated steps, each of which modifies a molecule slightly until the desired compound is produced. For example, the amino acid leucine is formed from pyruvic acid by a series of steps:

$$\text{Pyruvic acid} \xrightarrow[-CO_2]{\text{enzyme-1}} (a) \xrightarrow[\substack{+ \text{ active} \\ \text{2-carbon} \\ \text{compound}}]{\text{enzyme-2}} (b) \xrightarrow{\text{enzyme-3}} \text{Leucine}$$

Now suppose that leucine becomes scarce in the environment and that it is required for protein synthesis. Cells that happen to contain proteins that can modify compound b, which is still available, will be strongly selected. In like fashion, as compound b becomes scarce, those cells that can modify a to b and then b to leucine will be selected. This evolutionary pattern continues throughout the whole complex network of a cell's biochemical potentials, forced through the requirement of a common trophic level of ecology to develop on the basis of a common set of small organic compounds.

All biological systems use the same basic biochemistry— the same metabolic pathways for sugar utilization, for amino acid synthesis, for fat synthesis and degradation, for lipid synthesis, and so on. Ecology's common trophic level asks only that a common set of small molecules be used; it does not appear to require a universality of metabolic paths. This commonality of function can be explained in two ways. According to the first, all present life is descended from a single stem population. We might reject this approach since it uses historical accident to explain a desired result. The alternative is to say that every metabolic path in our present biochemistry uses to an evolving maximum those molecules that are uniquely suitable. At present this explanation rests upon the faith that order exists in the Universe. There is no way to test the tenet of suitability save to examine life of independent origin; and this is not available.

With the evolution of metabolic pathways, new environmental niches open at an expanding rate. Heterotrophic systems—those that use organic compounds as sources of chemical potential energy—arise, along with various other forms of metabolism.

Solar ultraviolet radiation provides energy, but it also damages cells. Protein structures become degraded and nonfunctional; nucleic acids strongly absorb ultraviolet radiation and are degraded. Porphyrins, whose postulated role we discussed earlier, absorb both ultraviolet radiation and visible light. At depths of several meters in ponds, most ultraviolet radiation is absorbed by the water itself. This is not so for visible light. One can imagine intense early selection for utilization of visible sunlight, which essentially requires chlorophylls (a porphyrin, which is present) and enzymatic electron-transport systems (which can be selected).

As photosynthetic cells using visible light appear, free oxygen is released into the atmosphere as a byproduct of photosynthesis. In time, biological oxygen production outruns use of oxygen in geological cycles. The ozone ultraviolet radiation shield of the Earth's upper atmosphere was formed some 2250 m yr ago. From then until about 600 m yr ago, the oxygen level gradually rose to its present 20% by volume in the atmosphere. As this occurred, possible ecological niches again expanded. When the free oxygen concentration exceeded 3%, its use as a highly efficient electron acceptor in aerobic metabolism became possible.

All these and many other events can happen once a genetic apparatus and a protein-synthesizing system are present to provide cell populations that natural selection can act upon. In fact, the origin of life is no longer a question once a cell possessing these systems appears.

CHAPTER TWELVE

THREADS

. . . the end of all our exploring
Will be to arrive where we started
And to know the place for the first time.

T. S. ELIOT

The first human footprint on the Moon. (Photo courtesy of NASA.)

*T*his portrayal of the origin of life has assumed a great deal and left out a great deal more. In part this has been done to keep us from losing sight of central issues. In part it has been to enable me to present my views rather than simply to review others', which have been covered quite thoroughly in numerous other books. At this point a number of loose ends, justifications, and reconsiderations remain.

First, many biological small molecules are optically active, or chiral. Chiral molecules are mirror images of one another, like our left and right hands (Figure 12-1). When amino acids are synthesized by nonbiological means, both kinds of enantiomers (mirror-image forms) of each amino acid are produced in equal abundance. Thus we find equal amounts of D-valine and L-valine, of D-leucine and L-leucine, and so on, when a spark discharge experiment is conducted or when amino acids are recovered from a meteorite. (The designations D and L refer to optical properties of pure forms of the enantiomers.)

Biological systems require specific enantiomers: they use only L amino acids in their proteins, only D-ribose in RNA, and so on. This is because the shape of biopolymers in space requires either all D or all L monomers. No protein could achieve the regular periodicity of structure it needs in order to be catalytic, were it to use some D and some L amino acids. RNA would not present a regularly spaced array of bases if both D and L forms of ribose were used for its backbone structure.

But why do we use only L amino acids, and only D sugars? The chemical properties of D and L enantiomers were long thought to be identical. If this were exactly true, then it would seem reasonable that we use the enantiomers we do purely because they were selected by chance. But it seems not to be true.

If you scan a handbook of chemistry and physics, you quickly see that the physical properties of some enantiomers

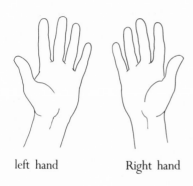

left hand   Right hand

L-alanine   D-alanine

FIGURE 12-1

Mirror image aspect of D and L enantiomers.

vary widely. Solubilities, melting points, and other properties are significantly different. As a consequence, we can state that in general the chiral molecules used in our biology were selected during early evolution for their slightly more advantageous properties.

W. Thiemann, in Germany, recently studied the polymerization rates of activated D and L amino acid mixtures. He found that the polymers formed always tend to contain more L than D amino acids. Other workers, as a curious aside to their main conclusions, have reported similar phenomena.

More study is required to discover exactly what chemical features of L amino acids make them more likely to be polymerized. Thiemann's observation strongly supports the contention that those chiral molecules chosen by biology were uniquely suited to their roles and were not selected by chance.

Throughout this book we have used the assumption of mediocrity. Made of the same material that occurs everywhere in our Universe, our planet formed from detritus about its protostar. Our planet, like any other in its area and of its size, outgassed its atmosphere, which then reacted and ultimately formed our biosphere. Wherever we study carefully, we find explanations: why we are constructed the way we are, why cells are the size they are, how a genetic system will develop. Our conclusion might well be that the Universe has one best and only way to make a biology, just as it has one best and only way to conduct hydrogen fusion.

But isn't this view really a subtle form of terrestrial chauvinism? Couldn't there be other equally good routes by which an ecosystem could develop besides the one we know? Selections operating during chemical evolution and natural selections operating on genetic protocells, blindly made over vast spans of time, represent evolutionary turning points that, once passed, are gone forever. If the time sequence of various selective events were different, might we too not be different?

At present these are unanswerable questions. We can make a very strong case that our set of bioelements is the only possible one and that the cell is the logical unit for any ecosystem. But we can find no compelling argument to force an ecosystem to produce a specific kind of tree, or us.

One step toward resolution of these uncertainties is to study life on other worlds which arose independently from similar events. The Viking mission to Mars reported no evidence of organic materials or of microbial metabolism—hence no life. Venus seems to be far too hot. Mercury has no atmosphere and is too close to the Sun. The giant planets, Jupiter and Saturn, are outside the realm of our discussion, if only because they are extremely different chemically and physically. We appear to be the only life that has originated within the solar system. To be alone is tragic. Even if our companions were Martian microbes their presence would provide one vital test of the soundness of our postulate that life is inevitable. Study of life forms of independent origin would allow us to assess the extent to which chance operates in selective processes.

To console ourselves for being alone in the solar system, we may ponder the notion that our galaxy has some 10^{11} stars, and that our models of star formation imply that planets are the inevitable outcome of star formation. But to prove that we are not alone we must go to the stars—or bring them to us.

CHAPTER THIRTEEN

REALITIES

We are children of the eighth day.
THORNTON WILDER

The Viking Lander. (Photo courtesy of NASA.)

This book has not given all the answers, since we don't know them all. Some "answers" have been presented as fact, some as very likely models, and others as plausible guesses that remain to be tested by future experiment. These different classes of answers have not always been delineated with the clarity a scientific monograph would require, since the principal object of this book has been a quite different one: to develop a coherent local cosmology. This chapter will point out those areas in which controversy and guesses exceed fact and will suggest some of the many kinds of experiments that need doing.

First of all, planets may not be the common outcome of the development of stars. Some astronomers—Shiv Kumar, for example—believe that almost all star systems tend to become binaries, most often of the type in which one star is relatively large and the other quite small. Indeed, if Jupiter were five times more massive than it is, it too would be large enough to begin internal nuclear reactions, and our system could be a binary, quite different from what it is. Although our argument in favor of the ubiquity of planetary systems is based upon reduced angular momentum of the central star of our system, the angular momentum problem could also be solved by a binary system in which the larger star has lost its angular momentum to its smaller companion. At present we cannot tell which school of thought is more nearly correct. The only definite observation we have is that of our own planetary system—we have no reason to believe that the forces that created it are unique in any way. Certainly, many aspects of our local cosmology will be on firmer ground when we can determine whether planetary systems are common or unusual.

Second, the presentation given here of the first primeval atmosphere is most unusual. Some researchers favor an atmosphere of methane and ammonia, basing their arguments on chemical equilibrium data. This aspect of the problem was discussed in Chapter Four. However, we have not discussed other possible atmospheres consisting mainly of carbon dioxide, water, and nitrogen, especially as proposed by W. W. Rubey. Essentially, Rubey's thesis is that if one balances the Earth's budget of carbon dioxide and other excess volatiles, one finds that carbon dioxide was probably a far more common gas in primeval days than now. This may be so; however, other facts do not seem to support this position. First, Rubey does not take into account the fact that carbon monoxide was at least five times more abundant than carbon dioxide in primeval volcanic emissions. Second, if considerable carbon dioxide were present, it would have dissolved into the ocean systems of the era. Excess carbon dioxide dissolved in water would tend to make the primitive ocean systems acidic. Weathering of rocks would have been extremely rapid, and the minerals deposited in sedimentary rocks during that era would have been quite different from rocks deposited later. This is not the case, however; no known sedimentary rocks hint at such discontinuities.

Third, our presentation of chemical evolution experiments has suggested that all biomonomers can easily be produced by suitable simulation experiments. Only in the most general sense is this true. Compounds such as sugars, which are quite chemically reactive, have not been recovered from general chemical evolution experiments. Various sugars can be synthesized from concentrated alkaline solutions of formaldehyde, and this has served as an argument for their general synthesis; however, none have been recovered from simulation experiments such as spark discharge reactions. This should not be surprising to the organic chemist, since any sugars synthesized would have quickly reacted with all manner of reactive intermediates and amino acids that are produced in abundance. Agreement is general that in the open, wide world of warm little ponds, sugars would have had many micro-niches in which to exist, but in the closed world of a crowded simulation flask their half-life would be too short for the chemist to recover and detect them before they would react with other compounds.

Another difficulty with chemical evolution resides in the "selection problem." What chemical mechanisms select specifically for those amino acids that constitute the biological subset? Why couldn't we have another set of purines and pyrimidines rather than the ones manifested by our biology? Few quantitative answers to these problems are available now, but we have hints that they are forthcoming. For example, recent research has found that complexes of clays with specific metallic ions will bind selectively to the biological subset of amino acids but not to others. Other chemical studies have shown that the purine and pyrimidine bases used in our biology are the most stable among that possible larger set which will still permit base pairing by hydrogen bonding. A more disturbing problem, unapproached at the moment, is the very unstable nature of sugar molecules and the central role they play in our entire metabolism. It appears logical that since sugars are such touchy, unstable molecules that they should be the cornerstone of our metabolism, but no convincing work has been done to show exactly why this might be so.

Did protocells really have to come first, or might clays, evaporites, and other mechanisms have provided sufficiently high local concentrations of polymers that metabolism could have originated before protocells? Occam's razor—the principle that the hypothesis requiring fewest statements is bound to be correct—suggests that the protocell-first argument is the most likely. Microfossil evidence supports this contention. Several groups of researchers are actively studying the effects of various clay minerals on the polymerization of amino acids and nucleotides. Are they wrong, then, and wasting their time? No. Whatever their immediate role in primeval times, clays (and other well-characterized surfaces) represent a two-dimensional surface analogous to the many surfaces within a protocell. Whatever we learn from the well-defined properties of clay surfaces can be applied with great profit to the many internal, phase-bounded surfaces of protocells. However, proposed concentrative mechanisms involving evaporates bear most careful scrutiny and perhaps experimental tests. When it is proposed that a pond containing amino acids evaporates to leave crystalline amino acids ready to undergo heat-driven anhydrous condensation reactions, the totality of the issue must be examined. Other molecules are present as well as amino acids. No one has yet evaporated the products of a spark discharge flask upon reconstituted sea water (for one example) and then subjected these contents to heat to see what might happen.

How did the genetic apparatus evolve? The mechanism proposed here is a guess. Any and all mechanisms proposed will be guesses until we have concrete data based upon analysis of extraterrestrial life. This is because the genetic apparatus is the product of close to 1000 m yr of evolution. Guessing over these time spans with uncertain environmental data and selection routes is near folly. The scheme presented here is novel in that a relationship based upon permutations necessarily exists between the discriminator and anticodon ends of primitive transfer RNA molecules. Novelty does not make right, and guesses such as these are only guesses. But this suggestion is more than a guess. A fixed relationship between protein and nucleic acid base sequence exists in our current biology. For this relationship to exist now, it must have existed in some way at the very origin of the system. Any mechanisms proposed for the origin of the genetic apparatus must face this issue.

The selection mechanism by which optically active forms of amino acids, sugars, and other molecules become incorporated into evolving prebiological systems at the expense of their enantiomeric forms has been presented as known. It may not be. The mechanism proposed was based on the observations that the physical properties of D and L enantiomers differ. Some researchers are not aware of these small differences; others maintain that the differences result from laboratory errors or from working with impure samples of D and L forms. With the exception of Thiemann's recent work, presented in Chapter Twelve, no one has directly searched for small differences in the physical properties of the enantiomers. This work sorely needs doing. Morowitz calculated in 1969 that small differences should exist but presented no experimental confirmation. As Thiemann's thorough review shows, the research literature abounds with examples that small differences exist. We await the crucial experiment.

The evolution of eucaryotic complex cells as a symbiosis of differently talented procaryotic cells has been glossed over quickly. This is not to indicate that this evolutionary progression is not of great and consuming interest. Rather, it is a different problem, one of early evolutionary biology.

Has evolution stopped with us? Are we what it was all for? To believe this is to retain, by analogy, the view that Earth is at the center of the Universe. Already at the infancy of our technological culture we can easily conceive of the notion—if not the possibility—of self-reproducing automata. Computers are permitted to enter chess tournaments as long as they restrict the time they use per move of the pieces. Active experimental work goes on concerning the development of artificial computer intelligence, to the extent that some machines and programs can now talk to the unwary beholder and demonstrate their essential human qualities. Perhaps we will serve as a race of smart monkeys who have designed even smarter computers to perform the real work of the Universe. Even so, we will leave our computers with a very human legacy, perhaps the most human of all—curiosity.

BIBLIOGRAPHY

This list of references is by no means comprehensive. A complete bibliography on chemical evolution and the origin of life is available:

West, M. W. and Ponnamperuma, C. "Bibliography on Chemical Evolution and the Origin of Life." *Space Life Sciences* 2:225–295, 1970.

This compilation contains more than 1500 references to both books and journal articles. The West/Ponnamperuma bibliography is updated periodically in issues of the journal *Origins of Life,* published by D. Riedel Company, Dordrecht, Holland.

Further Reading for Chapters One, Two, and Three

Faul, H. *Ages of Rocks, Planets and Stars.* New York: McGraw-Hill, 1966.

Gingerich, Owen, ed. *New Frontiers in Astronomy: Readings from Scientific American.* San Francisco: W. H. Freeman and Company, 1975.

Gingerich, Owen, ed. *Cosmology + 1: Readings from Scientific American.* San Francisco: W. H. Freeman and Company, 1977.

Hoyle, Fred. *Highlights in Astronomy.* San Francisco: W. H. Freeman and Company, 1975.

Hoyle, Fred. *Ten Faces of the Universe.* San Francisco: W. H. Freeman and Company, 1977.

Further Reading for Chapter Four

Clark, S. P., Jr., Turekian, K. K., and Grosman, L. "Model for the Early History of the Earth." In *The Nature of the Solid Earth,* edited by E. C. Robertson. New York: McGraw-Hill, 1972.

Gass, I. G., Smith, P. J., and Wilson, R. C. L. *Understanding the Earth.* New York: Academic Press, 1971.

Kummel, Bernard. *History of the Earth: An Introduction to Historical Geology,* 2nd edition. San Francisco: W. H. Freeman and Company, 1970.

Mason, Brian. *Principles of Geochemistry.* New York: John Wiley & Sons, 1966.

Phinney, R. A., *The History of the Earth's Crust.* Princeton, N.J.: Princeton University Press, 1968.

Press, F., and Siever, R., eds. *Planet Earth,* 2nd edition. San Francisco: W. H. Freeman and Company, 1978.

Urey, H. C. *The Planets.* Chicago: University of Chicago Press, 1952.

Urey, H. C., "Evidence Regarding the Origin of the Earth." *Geochimica et Cosmochimica Acta* 26:1—14, 1962.

Wilson, J. T., ed. *Continents Adrift and Continents Aground: Readings from Scientific American.* San Francisco: W. H. Freeman and Company, 1976.

Windley, B. F., ed. *The Early History of the Earth.* New York: John Wiley & Sons, 1976.

Further Reading for Chapters Five to Twelve

Bernal, J. D. *Origin of Life.* Cleveland: World Publishing Company, 1967.

The Biosphere: A Scientific American Book. San Francisco: W. H. Freeman and Company, 1970.

Buvét, R., and Ponnamperuma, C., eds. *Molecular Evolution I: Chemical Evolution and the Origin of Life.* Amsterdam: North-Holland, 1971.

Calvin, M. *Chemical Evolution.* Oxford and New York: Oxford University Press, 1969.

Ehrensvärd, G. *Life: Origin and Development.* Chicago: University of Chicago Press, 1962.

Fox, S. W., ed. *The Origin of Prebiological Systems.* New York: Academic Press, 1965.

Gibor, A., ed. *Conditions for Life: Readings from Scientific American.* San Francisco: W. H. Freeman and Company, 1976.

Haldane, J. B. S. *Science and Human Life.* New York: Harper and Brothers, 1929.

Hanawalt, P. C., and Haynes, R. H., eds. *Chemical Basis of Life, An Introduction to Molecular and Cell Biology: Readings from Scientific American.* San Francisco: W. H. Freeman and Company, 1973.

Keosian, J., *The Origin of Life.* New York: Reinhold Publishing Corporation, 1964.

Kimball, A. P., and Oro, J. *Prebiotic and Biochemical Evolution.* Amsterdam: North-Holland, 1971.

Margulis, L., ed. *The Origin of Life.* London: Gorden and Breach, 1970.

Oparin, A. I. *The Origin of Life.* London: Macmillan, 1936.

Oparin, A. I. *Genesis and Evolutionary Development of Life.* New York: Academic Press, 1968.

Rutten, M. G. *The Origin of Life by Natural Causes.* Amsterdam: Elsevier, 1971.

INDEX